Across Boundaries

Across Boundaries

Transborder Interaction
in Comparative
Perspective

Edited by

Oscar J. Martínez

and
The Center for Inter-American and Border Studies
The University of Texas at El Paso

First Edition
ISBN 0-87404-097-3 (paper)

Contents

Acknowledgements

On 21 March 1984, the Center for Inter-American and Border Studies at The University of Texas at El Paso hosted a conference on "Problem Solving along Borders: Comparative Perspectives." The papers presented at that meeting constitute the material for this book. The Andrew W. Mellon Foundation, through the U.S.-Mexico Border Research Program at The University of Texas at Austin, provided financial support for the conference and for the publication of this volume. I thank the late Stanley R. Ross, former director of the U.S.-Mexico Border Research Program, for facilitating the grant, and Howard G. Applegate, faculty and research associate of the Center for Inter-American and Border Studies, for his role in securing it.

I also wish to thank all the participants at the conference. My work was made easier by the timely submission of the essays and the excellent spirit of collegiality that prevailed during the entire project. The idea for the conference emerged from a conversation with Ivo Duchacek and Niles Hansen in the 1983 meeting of the Association of Borderland Scholars; I am grateful for their encouragement, support, and participation. Apart from the contributors to the volume, whose names appear elsewhere, I should like to acknowledge the participation of the following persons in the conference: Opening Remarks: Haskell Monroe, president, The University of Texas at El Paso; moderators: Kathleen A. Staudt, Department of Political Science, The University of Texas at El Paso, and Guillermina Valdez, Ciudad Juárez Office of Mexico's Colegio de la Frontera; commentators: C. Richard Bath, Department of Political Science, The University of Texas at El Paso, and George Rodríguez, Texas Governor's Border Regional Development Office, El Paso.

I express my appreciation to the staff of the Center for Inter-American and Border Studies, especially Larry McConville, Betty Manriquez, and Anne Holder, who provided editorial assistance, and

Rosalía Solórzano Torres, Rose Torres, and Jackie Dwyer, who contributed in a number of ways to the project. Thanks are also due the Board of Texas Western Press, and in particular its director, Dale Walker, for the publication of the manuscript.

Introduction

International borders present challenging situations at various levels of human interaction. The individual seeking to cross legally from one nation to another must satisfy immigration and customs requirements, which at times can be tedious and intimidating. It is common for border crossers to experience anxiety and to feel like anonymous entities at the moment questions about citizenship have to be answered and personal possessions have to be inspected. Difficulty clearing either of these two hurdles can result in delays, disrupted travel schedules, and great frustration. In the case of tourists and other short-term visitors, an initial unpleasant encounter with border functionaries may trigger negative feelings toward the country to be visited, whereas a quick and successful ingress tends to have the opposite effect.

Casual crossers, whose major objective is to go through inspection points without complications, seldom stop to think what a border means to a region or a country. Even residents of border towns who regularly commute from one side to the other often fail to appreciate many of the unique aspects of living in a border situation. Nevertheless, borderlanders usually have a general understanding that interaction with "the other side" is frequently a matter of necessity and even survival. This is especially true in cases where a pronounced level of economic interdependence exists. Ways must be devised to counteract the exclusionary features inherent in the boundary, not the least of which are government policies to limit contact between the two sides. Aided in many cases by common ethnic and linguistic bonds and spurred by regional self-interest, border populations may promote transboundary interaction, even if such behavior is frowned upon by the federal authorities of each nation. The latter, who are charged with controlling and administering the flow of people and goods across boundaries, attempt to regulate interchange to suit national interests; frontier needs and desires are of secondary importance. Commonly, regional interests clash with those of core areas, and therein lies the cause of perennial friction between many border areas and their respective national governments.

At the level of international relations, boundaries have always been a source of trouble, although the nature and degree of conflict have varied from border to border. Moreover, many nations have experienced significant evolutionary change in their international relationships, leading to different dynamics along border frontiers. In earlier times most serious boundary problems centered on territorial disputes and violations of national sovereignty. Borders typically functioned as buffer zones intended to diminish the chances for international strife. Given the centuries of warfare that yielded constant alterations of boundaries in continents like Europe, that utilization of borders seemed justified. Nations that lost territory as a result of war had ample reasons to keep a safe distance from aggressive neighbors. Few incentives existed to develop and populate border frontiers. That scenario has changed for many nations as a result of reduced warfare and increased integration of the world economy. What countless countries confront now are modern-day issues such as illegal migration, smuggling, air pollution, water contamination, and trade protectionism.

As territorial conflicts declined after World War II and as nations modernized their transportation and communications systems, international trade increased; interdependence between neighbors became a way of life. Border regions began to figure more prominently in labor exchanges, commercial transactions, and binational industrialization. In parts of Western Europe and along the U.S.-Mexico border, the levels of integration between adjoining border regions reached unprecedented proportions by the 1970s. Once sparsely populated frontier zones became poles of attraction for capital and labor, spurring regional growth and development integrally linked with the international economy. Small border communities blossomed into middle-sized and large cities, creating new opportunities for binational interchange.

While some people view these developments as positive, others have raised questions about the long-term consequences of pronounced integration and the implication of large population concentrations along border frontiers. Perceived threats to national security have rekindled nationalistic sentiments in some cases, governed by old notions that borders constitute "front lines" to be guarded at all cost from "undesirable" people and influences. Such attitudes, along with cultural chauvinism, insensitivity to local concerns, and ignorance of the need for transboundary cooperation have often impeded progress in solving problems between adjoining border communities that reflect fundamental changes in the global economy and have little to do with threats to national sovereignty. This unfortunate situation is illustrated particularly along the U.S.-Mexico border, where local residents have long fostered neighborly transborder cooperation without the degree of encouragement or assistance that is merited from the two respective federal governments.[1]

Like their counterparts in North America, for many years Europeans did little to promote transboundary cooperation. However, a new trend began in the early 1960s. A locally initiated effort to address a number of problems in the neighboring areas of Alsace (France), South Baden (Germany), and the border cantons of Switzerland led to the creation of Regio Basiliensis in 1963. This trinational body, which received the endorsement of the respective national governments, assessed the needs of the border area, collected data, and engaged in transportation and environmental planning. The emergence of Regio Basiliensis focused attention on the need for coordinated and institutionalized approaches to the solution of European transboundary issues.

By 1970, three associations of municipalities in the German-Netherlands frontier established the Euregio, a unique institution that includes a binational parliamentary council and a secretariat. The Euregio is committed to a comprehensive approach to border development rather than traditional piecemeal and separate efforts. Following these examples, most of the other Western European border areas have initiated cooperative ventures through a variety of institutional arrangements, most notably through organizations attached to such prominent bodies as the Council of Europe and the Liaison Office of European Regional Organizations. The mechanisms established to address European transboundary problems have resulted in significant progress for border populations and improved relations among frontier municipalities.[2]

The means by which binational communities in Europe have managed to make boundaries more permeable for the common good merits the attention of border regions around the world where greater transboundary cooperation and interchange is desirable. It is possible that arrangements formulated in the European context may be transferable elsewhere, allowing, of course, for appropriate modifications to suit local conditions.

The notion that the experiences of some borders might be of benefit to other borders, along with the broader idea that general comparisons of transboundary interaction in different geographic locations is in itself useful, were major considerations for holding the conference that led to the assembling of this volume. Under the co-sponsorship of the Center for Inter-American and Border Studies at The University of Texas at El Paso and the U.S.-Mexico Border Research Program at The University of Texas at Austin, scholars from three continents gathered at The University of Texas at El Paso on 21 March 1984, to discuss border phenomena. The first conference of its kind, it was, in the view of the participants and observers, a highly stimulating and enriching experience.

This volume consists of six parts. In Part I, Ivo D. Duchacek examines the trend toward greater participation of local and provincial governments in international affairs. His analysis of "global microdiplomacy" and "transborder regionalism" in the context of international interdependence provides us with a helpful framework for viewing border interaction in Europe, Africa, and North America.

Part II includes essays by Niles Hansen and Hans Briner. Hansen points out that the Western European borders, as well the U.S.-Mexico border, are no longer isolated and underdeveloped regions but are highly integrated areas that have many of the economic characteristics of core areas. Former disadvantages posed by the presence of borders have been overcome through improved international relations, increased interdependence, and greater cooperative efforts at the local level. Hansen argues that many of the conditions that spawned transboundary cooperation in Europe, especially in the France-West Germany-Switzerland region, also exist along the U.S.-Mexico border. Thus, it is in the interest of Americans and Mexicans to look closely at the Western European models. Briner elaborates on Hansen's observations pertaining to the Regio Basiliensis by describing in some detail how transfrontier planning has progressed despite fundamental dissimilarities in local autonomy and restrictions imposed by different federal jurisdictions.

Part III focuses on problem solving along the U.S.-Mexico border. Ellwyn R. Stoddard points up the need for developing a realistic frame of reference if efforts to address border problems are to be successful. Federal authorities, especially in the United States, tend to view the boundary as a barrier whose supreme function is to protect national sovereignty. That perception ignores the existing intense economic and cultural interpenetration as well as the overlapping of myriad problems across the border. Stoddard, arguing that the problems of one side are also the problems of the other, proposes the acceptance of a "Doctrine of Mutual Necessity," whereby both sides recognize the existing integration, put aside nationalistic impediments, and tackle issues in a binational framework. Gustavo del Castillo is likewise critical of policymakers who fail to see border problem solving as a bilateral matter. Del Castillo focuses on recent regulatory actions of governmental structures to illustrate the slow, complicated, frustrating, and ineffective approaches to dealing with border issues. His examples include the devastating impact of peso devaluations on the border communities and environmental contamination in the Tijuana-San Diego area. In his essay, Lawrence A. Herzog presents a conceptual model for understanding the transboundary urban ecosystems that have developed along the U.S.-Mexico border. Herzog demonstrates the symbiotic character of the San Diego-Tijuana ecosystem and underscores the urgency of dealing with a host of shared problems. He identifies serious weaknesses in the current methods used for reaching solutions and calls for the creation of effective transboundary mechanisms for cooperative management of the border.

Part IV consists of two essays on the U.S.-Canadian border. Donald K. Alper discusses the increasing interaction between the semi-autonomous Canadian provinces and their counterpart U.S. states. Because both national governments tend to ignore transboundary problems, provincial and state officials have felt a need to deal directly with each other. One result has been the creation of border-focused institutions at the state level, as in the case of Maine, to strengthen the capacity for handling situations that transcend the boundary. Victor Konrad discusses transborder cultural transfers, with an emphasis on Canada's influences on the United States. He uses the spread of the Madawaska Twin Barn as a prime example of diffusion and adaptation.

In Part V, Anthony I. Asiwaju discusses border problem solving along the Nigerian-Benin frontier. He stresses that recognition of the ethnic coherence and interlink among the three cultures that inhabit the borderlands is essential to understanding why local people have a sense of rejection of the boundary as a dividing line, although they recognize it serves to divide French and English institutions and traditions that remain from colonial days. Asiwaju's comparative comments between this fascinating African border and the U.S.-Mexican border

6 • ACROSS BOUNDARIES

are particularly instructive, demonstrating the need for viewing trans-boundary interaction in a universal framework.

The volume concludes (Part VI) with Z.A. Kruszewski's examination of how some Communist borders function. Kruszewski notes that in recent years transborder interaction among several Eastern European nations has significantly increased owing to a relaxation of once rigid rules pertaining to international contact and exchange. Central to the now greater fluidity of those borders has been the emergence of tourism as an important economic activity. Kruszewski's observations concerning migratory flows, trade patterns, and smuggling underscore the proposition that, irrespective of political systems, a universal border behavior exists.

Although the essays vary in their content, they are unified by the concerns of all the authors with the nature of border interaction and the approaches to dealing with binational issues in a regional context. The case studies make it clear that many borderlands have become highly integrated, which situation has significant implications for the respective nation-states. New approaches to border problem solving are necessary. A logical start is for national leaders to recognize the universality of many of the problems that confront them at their frontiers. The realization that "we are not the only ones that have these problems" might facilitate establishing realistic frameworks within which efforts may be made to find solutions. Additionally, decision makers at all levels representing different countries need to work closer addressing border issues.

For scholars, these essays suggest the imperative for research in comparative border studies. There is considerable work in progress on individual borders around the world, but far too little has been done to link these studies and to begin constructing a theory of border behavior. Specialists in the U.S.-Mexico border in particular have neglected delving into the broader implications of their work.[3] By identifying some common patterns of border interaction and by suggesting concepts and approaches to further our understanding of border phenomena, the essayists in this volume have made a significant contribution to the field. Comparative border research remains largely unexplored, however. Interested scholars have a challenging agenda ahead of them.

Oscar J. Martínez
El Paso, 1 June 1986

Notes

1. The emergence of informal arrangements along the U.S.-Mexico border to circumvent national statutes and policies is discussed in: John W. Sloan and Jonathan P. West, "Community Integration and Policies among Elites in Two Border Cities: Los Dos Laredo," *Journal of Inter-American Studies and World Affairs* 18 (November 1976): 451-71; John W. Sloan and Jonathan P. West, "The Role of Informal Policy-Making in U.S.-Mexico Border Cities," *Social Science Quarterly* 58 (September 1977): 270-82; and Ellwyn R. Stoddard, Oscar J. Martínez, and Miguel Angel Martínez Jasso, *El Paso-Ciudad Juárez Relations and the 'Tortilla Curtain': A Study of Local Adaptation to Federal Border Policies* (El Paso: El Paso Council on the Arts and Humanities, September, 1979).

2. Niles Hansen, "European Transboundary Cooperation and Its Relevance to the United States-Mexico Border," *APA Journal* (Summer, 1983): 339-41.

3. The reluctance of U.S.-Mexico border scholars to venture beyond the narrow confines of their geographical region is aptly discussed in Anthony I. Asiwaju, *Borderlands Research: A Comparative Perspective* (El Paso: Center for Inter-American and Border Studies, The University of Texas at El Paso, Border Perspectives Paper No. 6, November 1983).

Part I

Non-Central Governments and Transborder Diplomacy

International Competence of Subnational Governments: Borderlands and Beyond

Ivo D. Duchacek

The meaner sort of men confine their outlook within the cities where they were born. But those whom God has given a greater light neglect no means of improvement whether it comes from near or from afar. . . . The light of nature teaches each of us in his private life to maintain relations with his neighbors because as their near presence enables them to injure so it also enables them to do us service.

Cardinal de Richelieu (1585-1642)

Can your city afford to continue ignoring the rest of the world? Not if your local economy is going to grow. . . . Maybe it's time to expand your city's horizons.

The National League of U.S. Cities,
International Trade Development
Conference, San Antonio, Texas,
(9-11 September 1983)

As managers of economic, social, and cultural affairs, noncentral (subnational) governments have become increasingly sensitive to pressures as well as to lures of global and regional interdependence. In other words, they directly experience their vulnerability to events abroad beyond their own and their nation's control — as in the case of the OPEC oil embargo and subsequent world recession. For them, too, "dependence and interdependence are not different concepts but different locations on the same conceptual continuum."[1]

As a result, noncentral governments have willy-nilly appeared on the international scene alongside the national (central) governments as coactors in international relations. Every so often they complement, but also compete with, their central governments in that portion of national foreign policy dealing with economic, social, ecological, labor, humanitarian, and cultural issues. Patricularly is this confounding of jurisdictions apparent in the case of significantly decentralized unitary systems from which subnational territorial autonomy, as it were, tends to ramify into international politics — until recently a seemingly "eminent domain" of central governments. Thus, paradoxically, provincialism has been internationalized and localism globalized.

The present rapid and mass movement of products, persons, power, and pollutants necessarily affects some areas and their governments within a nation more than the other territorial components of the nation-state or the nation as a whole. Acid rains affect the U.S. Northeast and Eastern Canada much more than the West; the problem of legal and illegal crossings into the U.S. Southwest is a major one in Southern California and Texas but less so in Montana and New York; warming up of the Rhine waters by an excessive number of nuclear reactors in the Rhine corridor alarms French Alsace, northern cantons of Switzerland, and south Baden-Württemberg considerably more than the French Riviera, Ticino, and Hamburg. As could be expected in our era of interdependence, both central and noncentral governments feel impelled to react to the various external threats and opportunities to satisfy their respective electorates' needs and so stay in power; this is one of the consequences of modern government ambition and endeavor to manage economic and social affairs. Our emphasis here is that presently *horizontal* cooperation across national borders to maintain governmental controls *vertically*[2] characterizes not only central but noncentral governments as well.

Subnational elected officials, their staffs, and their publics have asserted their international competence only in those areas they perceive as being basically within their local/provincial/state/cantonal jurisdiction — whether reserved, residual, shared, or traditionally delegated. Except in the case of a secessionist region that aspires to independent sovereign statehood, noncentral governments continue to recognize that all-national diplomatic status and defense remain within the monopolistic embrace of the central government. The problem naturally arises that many an economic, technological, humanitarian, or ecological issue has some national security aspects. Likewise is the reverse true, so that security issues have provincial/regional implications, as, for example, in the case of placing missile silos in Utah and Nevada in 1982. The line between security concerns, on the one hand, and nonsecurity issues is obviously quite fluid and certainly not as clear as the authors of old or new constitutions and traditional diplomats seem to believe. Thus the internal and external workings of federal and decentralized systems have been and will continue to be affected by the international activities initiated by noncentral governments. This is one of the main foci of this study.

The projection of subnational needs and interests directly onto the international scene has so far assumed two distinct, though quite often interlaced forms: *global microdiplomacy* (or *paradiplomacy*) and *transborder regionalism*. The first reflects the subnational awareness of global interdependence, whereas transborder regionalism represents a geographically circumscribed response to the risks and opportunities for cooperation arising from contiguity. As Richelieu expressed it nearly four hundred years ago, neighbors can injure as well as do us service. Contemporarily, it is the spatial dimension — worldwide as opposed to regional — which differentiates the two closely related forms of the ramifications of subnational territorial autonomy. Another difference is that more often than not in global microdiplomacy, noncentral governments are in direct contact with the functional sectors of the foreign central government (such as department of commerce) besides private citizens and corporations; in the transborder regional framework noncentral governments deal with their noncentral counterparts on the other side of the frontier. Often two noncentral transfrontiers exercise coordinated pressure on their respective central governments or on centrally appointed and managed border institutions. A new study on the transborder regional influence of the International Boundary and Water Commission in the Southwest illustrates well this process of "regionalization" of a federal institution.[3]

In both cases of regional and global microdiplomacy, under the pressure of scarcity, the eyes of noncentral governments, though still primarily directed toward the national capital and its financial resources, have also turned their sights outward, across the national boundaries toward not only their neighbors but also toward their neighbors' neighbors and distant corners of our interdependent world. Their autonomous life depends on their ability to establish mutually advantageous links with foreign sources of economic, financial, and technological power and foreign human resources. Both global paradiplomacy and transborder regionalism also represent a subnational negative reaction to big central government and its extension of national foreign policy monopoly into those areas which have usually been considered as falling under local/provincial jurisdiction. Since World War I central governments increased their proclivity to "legislate" in these areas by means of international treaties and executive agreements. Microdiplomacy, it may be said, fights back.

Global Microdiplomacy

The term global "microdiplomacy" or "paradiplomacy," as used here, refers to processes and networks through which subnational governments search for and establish cooperative contacts and compacts on a *global* scale, usually with foreign central governments and private enterprises. The aims of such subnational activities beyond the national borders are mostly export trade, reverse investment, cultural exchanges, environmental protection, and tourism as major sources of income. The actual forms of global microdiplomacy in which local and regional governments in North America engage range from short-term fact-finding missions abroad to promotional trips undertaken by subnational leaders (Canadian premiers, mayors of major cities, and U.S. state governors), to hosting foreign dignitaries and corporate or business leaders, and to establishing foreign trade zones (30 U.S. states have a total of 55 such zones).[4] Other microdiplomatic activities include trade and investment shows featuring local and provincial/state manufacturing and technological skills (for example, industrial shows such as "Invest in U.S. Cities" in Zurich and Hong Kong). Still other endeavors by subnational governments are special interprovincial relationships, linking up one town or province with another overseas (such as Alberta's special relationship with the Japanese island province of Hokkaido, the South Korean province of Gangweon, and the Chinese province of Helungjian), and — the most conspicious and characteristic feature of global microdiplomacy — permanent "paradiplomatic" offices in major foreign centers of political, financial, and economic power. Since the 1970s, twenty-eight U.S. states, for example, have established fifty-four permanent state offices in seventeen foreign countries; seventeen of these offices are in Tokyo and eleven in Brussels. In addition, eighteen U.S. ports have their representatives in major west European cities and ports.

Six of the ten Canadian provinces have created forty-six missions in eleven countries. Today twenty Canadian provincial paraconsulates operate in major U.S. cities in cooperation but also in competition with each other, as well as with the all-Canadian embassies and consulates. The Quebec delegation in Paris has eighty employees, as medium-sized embassies usually have. The province of Ontario, on the other hand, maintains a relatively small mission in Paris, comparable in size and function to U.S. foreign offices abroad.[5]

Global paradiplomacy as practiced by the two North American federal unions (but not Mexico) is a relatively new phenomenon, although some (mostly imperial-colonial) precedents existed in the past, such as Quebec representation in Paris since the seventeenth century and the intra-imperial and intra-Commonwealth paradiplomacy among the British earlier in this century. The vigorous pursuit of paradiplomatic possibilities by various U.S. states and Canadian provinces certainly reflects the traditional broad autonomy of the U.S. states in the U.S. federal system and even broader autonomy and power over natural resources enjoyed by the ten Canadian provinces for which we have found so far no real counterparts in Mexico, Western Europe, and other continents.

The emergence of the North American global version of microdiplomacy clearly dates back to the oil crisis of 1973 and subsequent economic recession which so painfully demonstrated the subnational vulnerability to distant changes and to the resulting decline of central financial support for the various developmental and welfare programs needed on the various subnational levels. In the U.S. the reasons for paradiplomacy in distant places are frequently referred to by noncentral governments quite openly and bluntly. The state governors, for instance, proclaimed in 1981:

> States must maintain and enrich their contacts with foreign governments, industries, and sciences. . . . They must be diligent in . . . establishing the need for recognition of their individual and collective interests in international affairs. . . . This explosion in the states' international activities reflects a growing awareness among governors of the degree to which foreign business trends and opportunities can affect the economic health of their communities.[6]

The chairman of the Advisory Commission on Intergovernmental Relations (ACIR), Arizona Governor Bruce Babbit, expressed the need for global paradiplomacy in 1983 even more matter-of-factly:

States will probably have to assert themselves as never before in modern times. . . . The message [of the fiscal pressure on the federal budget] is clear: The national government will no longer bail us out.[7]

Insofar as these paradiplomatic initiatives taken by state/provincial capitals complement rather than conflict with the conduct of national economic foreign policy, they have often been encouraged (or at least benevolently coordinated) by the central government — although Ottawa is obviously much more careful than Washington, D.C., in this respect on account of Quebec. While until 1984 the *Parti québecois* was committed to separatism under Trudeau, Ottawa viewed Quebec's paradiplomacy as protodiplomacy and its offices abroad as protoembassies of a would-be sovereign state. Between 1966 and 1984, Quebec often quite openly grafted the separatist program of the independentist *Parti québecois* onto its otherwise justifiable promotion of its autonomous economic and cultural interests abroad.

Even without any secessionist message, some subnational paradiplomatic actions are resented and opposed by the central government. In another study Duchacek[8] listed eight central objections to noncentral international activities:

1. Opposition in principle, invoking the constitution and the seemingly clear separation of foreign policy from all other policy.
2. Reluctance to increased diffusion of power on the part of those who hold it.
3. Fear of anything novel, threatening the bureaucratic routines and inertia.
4. Devotion to "institutional tidiness" that abhors new, complex, and complicated procedures, as well as the related potential for chaos.
5. Fear of foreign exploitation and misuses of provincial autonomy which, with some external help, may be asserted at the expense of other territorial components of the national system as a whole.
6. A related fear of the relative lack of skill and experience on the part of provincial and local staffs unfamiliar with negotiations procedures in contacts among sovereign nations.
7. Opposition to those state practices that could adversely affect the authority and international commitment of the central government (for example, its commitment to

GATT). In 1984 one such explosive issue was the case of the so-called "unitary tax" that some states imposed on transnational corporations. In the eyes of the U.S. Department of Commerce, foreign governments, and transnational corporations, the taxing was excessive.

8. Finally, the central government sometimes suspects that U.S. states may establish paradiplomatic offices abroad only as a status symbol — a "me-tooism" of sorts — thus creating only some minor problems of diplomatic courtesy and protocol for the central government rather than any problem of substance.

It should be recognized that in our missile age, the everpresent potential for nuclear holocaust (which would have little regard for both national and intrafederal boundaries) tends to project noncentral governments in free systems onto the international scene in still another way: in both the United States and Canada, but also in some West European towns, provinces, and districts, local authorities have become involved in organizing and passing various forms of resolutions in the framework of the anti-nuclear freeze movement. In the United States, for example, 11 state legislatures, 450 town meetings, and about 400 city councils passed such resolutions, probably with the hope that local majorities would thus significantly influence the national strategy of deterrence. It is a rather rare case in which local majoritarian victories have been used in national security matters, in contrast to the past when the national government usually tried to determine what local leaders and publics should think about international politics. It is interesting to note that some of the New Left groups, even though in principle devoted to the idea of socialist single party centralism, have adopted the tactic of mobilizing local majorities for particular ideological national and international goals.[9]

Transborder Regionalism

The concept of "transborder or transfrontier regionalism," as used here refers to the sum of the various informal and formal networks of communications and problem-solving mechanisms which bring *contiguous* subnational territorial communities into decisional dyads or triads — that is, bicommunal or tricommunal transfrontier regimes; in these an integrative transborder political culture may develop. Examples of such formal and informal transfrontier subnational regimes are regional frameworks linking West Germany and Swiss federal components with regions of the now decentralizing France, the Canadian and U.S. provinces in the Northeast and partly also in the West, and the numerous Mexican-U.S. informal networks in the Southwest. While for the time being our focus is on land contiguity, we recognize that "salt-water" or "fresh-water" neighborhoods should not be excluded from our consideration. Out of the ten Canadian provinces, four are not contiguous on land: Ontario is mostly a "fresh-water" neighbor of the eight U.S. states south of its border, while Newfoundland, Nova Scotia, and Price Edward Island represent "salt-water" U.S.-Canadian neighborhoods.[10]

A transborder *regime* is a set of rules and institutions, formal and informal, that aim at and succeed in regularizing neighborhood behavior. In Krasner's more general terms, a regime is defined as a "set of procedures around which actors' expectations converge."[11] Our study has transferred the concept of international regimes to geographically delineated regions that cooperate across an intersovereign boundary but do not aim at such integration as would give birth to a new "transfrontier nation," emotionally and structurally separate from the two respective political heartlands.

In the framework of the study of comparative federalism, cooperative transfrontier overlaps between two neighboring noncentralized or decentralized systems represent what may be called a confederal consociation of subnations. As is characteristic for all confederations, the aim of such cofederal regimes is an intersovereign cooperation without an intention to dissolve the basic link with and political allegiance to the respective national systems. If a peripheral province or state aims at secession (Quebec, for example), its goal is, of course, severance of the existing links and accession to the international system of sovereign states, not accession to a neighboring national system. As is characteristic of all consociational frameworks, decisions are reached by "amicable agreement" or near-unanimous consensus.[12] Hence do we derive our terms, global or regional microdiplomacy.

The Westphalian Birth Certificate of Transborder Cooperation

While the birthday of vigorous global paradiplomacy in North America may be roughly placed at the time of the OPEC crisis in 1973, no date can be identified for the emergence of transborder regionalism.

One may argue that transfrontier interactions are, in fact, as old as humanity, or, at least, as old as the first delineation between cavemen settlements for the purpose of determining the boundaries of their respective hunting grounds. The human need to organize life, ensure security, divide labor, maintain kinship, and promote collective identity within some identifiable space has led not only to territoriality; it has led to the simultaneous need to perforate it in order to ensure exit from and entry into the territorial corral — in principle supposedly impermeable. In this perspective even the most "natural" boundary must be perceived as basically artificial and permeable since it may conflict with human activities, inventions — starting with the wheel and the resulting mobility — or natural catastrophes such as the forest fire with its total disregard for political boundaries. (Provincial/state agreements concerning forest fires, flooding, and civil defense emergencies along the U.S.-Canadian borders are good examples.) It may be recalled that at its "ironest," even the so-called Iron Curtain was never impermeable: tape recorders, blue jeans, jazz and rock music, and Voice of America had little trouble changing Stalinist walls into sieves.

Most historians trace modern state territoriality — as well as its quasi-simultaneous permeability — back to the Peace of Westphalia, which in 1648 had fragmented Europe into sovereign corrals along religious and dynastic lines, even though these were soon modified by new conquests or peaceful, interdynastic arrangements. The European territorial divisions were later exported and projected on all other continents in the form of imperial colonialism and other "manifest destinies." Subsequently, these boundaries were challenged, modified, and perforated by industrial progress, technological innovation, transportation revolution, and mass migration. Rivers like the Rio Grande and the Rhine, viewed in one age as natural and convenient lines of separation, became in another age lines of contacts. Since fertile river valleys generally attract settlers from both sides of the river, they are, in a way, predisposed to become either sources of new conflict or new cooperation. The Rhine is a particularly interesting example of such a dual role. In many ways and in this historical framework, subnational cooperation across international boundaries quite often represents a rational corrective to many a previous territorial insanity.

Additionally, central governments may not only encourage and coordinate, but also initiate, and for their central purposes, manipulate the global and regional network of communication and negotiation among noncentral governments. Partly to defuse the separatist potential in Quebec's transborder regional cooperation in the Northeast, Canada's ambassador to Washington, Allan Gottlieb, during his annual visit to Toronto, Ontario, on 14 October 1983, recommended that "Ontario develop contacts with U.S. border state governors." So far Ontario has had a rather tepid attitude toward transborder regionalism, considering that its eight U.S. neighbors are all more or less competitive in the matter of export trade.

A very special case of central as well authoritarian manipulation of global and regional paradiplomacy is the chain of economic, semicapitalist zones established by the People's Republic of China in its coastal provinces, authorized to establish and maintain various joint enterprises with the capitalist West. So far, only one of the four authorized special economic zones has been impressively successful in attracting foreign high technology and business (Shenzen in Guangdong province near Hong Kong). The other special economic zones along China's coast facing the Pacific are Zhuhai and Shanton in the Guangdong province, and Xiamen in the Fujian province, opposite Taiwan and in clear competition with it.

To sum up at this point:

While it is useful to distinguish global paradiplomacy from geographically circumscribed transfrontier regionalism in terms of geographic expanse and central and noncentral partners, one qualification seems in order. It concerns the very concept of contiguity in our contemporary world with its supersonic speed of transportation and the mass movement of products, persons, and pollutants that makes close neighbors of us in whichever "global village" we live. The insistence of seventeen U.S. states to be represented in distant Tokyo, or Alberta's intimacy with Hokkaido, which now prefers Alberta's steaks above all others, are good illustrations. Nevertheless, as is well documented by the work of U.S. and Mexican borderland scholars, and the progress of Regio Basiliensis in the Alemanic Corridor along the elbow of the River Rhine, a close physical contact on a daily and mass basis still represents a decisive and historical ingredient in the process of transforming physical proximity into political intimacy.

Physical proximity permits easy and inexpensive communication on a daily basis. It is significant that most of the transborder cooperative business in our three research target areas (U.S./Mexico, U.S./Canada, and Germany/Switzerland/France) more often than not is conducted by telephone and luncheon appointments on both sides of the border.

It is also symptomatic that those U.S. states that maintain representative offices in distant places rarely have subnational paraconsulates in the neighboring Canadian provinces or Mexican *estados*. The Canadian provinces maintain their paraembassies in New York, Los Angeles, Houston, and San Francisco, not just across the borders in Albany, Seattle, Helena, or Montpelier. The Mexican *estados* apparently do not engage in global paradiplomacy at all, perhaps to be expected in a one-party federal system. Contiguous areas just across the border are simply not perceived as being "really abroad." The practice of transborder regionalism is primarily telephonic with a minimum of paperwork; it therefore poses some problems of monitoring for both the central government and the academic researcher.

How Subnations Negotiate

If by diplomatic negotiations we mean processes by which governments relate their conflicting interests to their common ones, there is, conceptually, no real difference between transborder regional and global paradiplomatic networks of communication and negotiation. In both cases the goal of regional or global microdiplomacy is the same as that of center-to-center macrodiplomacy: an *agreement* based on conditional mutuality. Both sides pledge a certain mode of future behavior on condition that the opposite side act in accordance with its promise. In contrast to domestic law, in international relations (whether on a micro- or macrodiplomatic level) no common superior authority can ever be invoked in case of violation.

Yet as we know from international politics, such unenforceable bargains are generally observed since both sides continue to have a very similar interest in (a) preserving the assumed advantage assured by the initial bargain; (b) adhering to one's pledge as a reaffirmation of credibility, an essential ingredient for future bargains; and (c) reconfirming the principle of good will.

The question to be raised in this otherwise self-evident framework is the following: How do the subnational authorities (that is, elected provincial/state/cantonal officials, their experts, and administrative staffs) respond to and operate in a basically anarchic milieu, governmentless in the sense of a total absence of central enforcement, in which conflicting interests may be related to common interests only by a negotiated agreement and its rational though unenforceable implementation?

How much of the hierarchical and majoritarian decision-making modes in which these administrators operate in the morning carry across the border in the afternoon when decisions may be reached only by consociational consensus ("talk until you agree")? In other words, deliberately imitating here a Harvard study on diplomacy, how do, in fact, *subnations negotiate?*[13]

We do not really know. Guessing, rather than empirical data based on analysis, guides us here. At one point, Stephen P. Mumme and this author considered the feasibility of a research project which would focus on subnational microdiplomacy. The results could have served both the Foreign Service Institute and the U.S. Pearson Program, which keeps sending about twenty-five foreign service officers for training in and exposure to international and transborder practices of state governments. Mumme and I were perhaps influenced by what a French diplomat had to say more than 250 years ago about necessary diplomatic skill in general:

> The diplomatic genius is born, not made. But there are many qualities which may be developed with practice and the greater part of the necessary knowledge can only be acquired by constant application to the subject. . . . When . . . a state is powerful enough to dictate to his neighbours, the art of negotiation loses its value. . . . The great secret of negotiation is to bring out prominently the common advantage that they may *appear* equally balanced to both sides. . . . (A negotiator should possess) an equable humour, a tranquil and patient nature, always ready to listen with attention to those whom he meets.[14]

Do the U.S. governors, French prefects, or Canadian premiers — our subnational "princes" — and their respective staffs have equable humor and tranquil and patient natures as the French diplomat required in 1716 of his aristocratic princes? Maybe we should find out and, if necessary, offer remedial courses in this field.

As perceived by the national center, an opposite deficiency or error may well need correcting: an excessive borderland "chumminess" as a reaction against the two distant capitals whose border policies have to be absorbed by the borderlands although they are policies "formulated far from the border by people unfamiliar with 'frontier customs' and regional integrative systems through which daily border activities are conducted."[15] One federal officer, commenting on undocumented Mexican maids in El Paso, expressed his general reaction to the author succinctly: "What's good for the border is not necessarily good for the other forty-six states. That's the problem."

In studying the borderland negotiating techniques in general, one has to consider that technical expertise, statistical data, and functional/professional interest groups seem to play a more important role in microdiplomacy than in traditional center-to-center diplomatic negotiations. Not only national political culture, but also national professional and transborder professional culture, endow transborder diplomacy with a different tone and different possibilities than is the case in macrodiplomatic negotiations — although even there transgovernmental professional/expert intimacy close to the apex may develop.[16]

Two Concepts in Search of Pre-Theories

Transborder regionalism and global microdiplomacy are concepts that still remain to be correlated properly to the workings and crises of national and international systems. Our ability to provide these relationships with explanatory coherence are at present outpaced by new developments in domestic and international politics that attract noncentral governments to play significant, though clearly secondary, roles on the international scene.

One obvious difficulty is our habit of viewing the nation states — speaking with one legitimate voice to the other nation states — as the primary units of analysis in both international and comparative politics. At the opposite pole, although more rarely, is the effort to explain national behavior by international systemic constraints and impulses. What is needed are improved theoretical tools to analyze, correlate, and explain the regional and global paradiplomacy in which so many noncentral governments now engage.

The initial search for a more suitable analytical framework has so far proceeded along the lines of the Nye/Keohane "world paradigm."[17] Our study has expanded and changed the paradigm, however, by emphasizing the opposition party and noncentral governments in addition to the Nye/Keohane list that consists of such nonstate actors as transnational corporations, ideological movements, churches, and international pressure groups (labor, managers, professionals, artists and intellectuals). I suggest the concept of a nation state as a multivocal (polyphonic) actor which, on the international scene, speaks with more than one central-government legitimate voice: the audible voices of noncentral governments can be heard, as well as the often strident voice of the potential government of tomorrow, the opposition party or the ethnic community aiming at independence. Diagramatically, the nation state *qua* multivocal actor could be illustrated in the form of a stepped Saqqara pyramid, with its separate yet interconnected points of entry on the international scene, in contrast to the neat, single apex of the Cheops pyramid.

It is evident that global and regional microdiplomacy poses some research questions which neither our research data nor our present theoretical tools can answer. A set of additional concepts and pre-theories is required to introduce some order for the purpose of explaining paradiplomatic activities, especially their causes and consequences. Some of the research questions and potential hypotheses are identified below, with the hope that further study may illuminate them:

1. What are the conditions causing subnational, autonomous authorities to feel compelled, short of secession, to search for cooperative contacts, compacts, and institutions to operate across the national boundaries? What are the options?

2. What are the effects of regional and global microdiplomacy on the internal and external — normal and abnormal — workings of democratic (noncentralized and decentralized) systems? What, in particular, are the constitutional, administrative, and coordinative changes that are imperative or useful? What could/should be the new channels of consultation between the nation and its territorial components?

3. What are the effects of regional and global microdiplomacy on international politics in general and international organizations in particular?

4. In what way does regional and global microdiplomacy contribute to cooperation among nations; in what way does it add new irritants to politics among nations?

In this context two research questions impose themselves:

A. If nation-to-nation interaction (on a center-to-center basis) increases, does the transborder regional interaction also necessarily increase (mirroring the emergence of good will on the international level)? Or does it, on the contrary, decrease, since "all is well" between the two nations and the border cooperation loses its significance?

B. If the center-to-center interaction and cooperation significantly decline, does interaction at the border also decrease, since local/provincial transborder issues cannot be really separated from the center-to-center alienation? Or does transborder paradiplomacy, in a reactive move, increase in volume to blunt the sharp edges of the center-to-center enmity? After all, ever so often the center-to-center tensions hurt the border relationship most painfully.

Scholars and politicians have voiced conflicting opinions on this last point:

A major finding is that the problems of the frontier zone are indeed an aggravated microcosm of the general issues dominating the relations between the two states — Mexico and the United States.[18]
The key to a new era of inter-American cooperation and trade will be found at the U.S.-Mexican border.[19]
The more the relations become tense between Ottawa and Washington, the more important it is that the Canadian provinces and the U.S. keep talking to each other.[20]

In Western Europe some scholars have argued that transborder regionalism is either an alternative to or a precondition for the unification of Europe. One of them, Charles Ricq of Geneva, says with apparent conviction:

Unlike the closed system of nation states . . . the transfrontier regions, transforming themselves from regions of confrontation to regions of concentration, will see the first real signs of surrender of sovereignty — Europe of the future will have to base itself on these regions in order to redesign itself and produce a better structure.[21]

In a more specific way, commenting on the nascent Alpine Community, consisting of parts of Switzerland, West Germany, Austria, Italy, and communist Yugoslavia, Ricq concludes, "The Community of the Alpine regions will necessarily contribute a foundation stone to the slow and complex construction of Europe."[22]

Is this a scholar's wishful thinking combined with obvious impatience with the slow progress of European unification? Or is it a realistic, modern estimate of the future roles of transfrontier regionalism whose relative symmetry may somewhat alleviate the gross asymmetry between powerful and powerless neighbors?

When we think of the U.S.-Canadian, U.S.-Mexican, and French/German/Swiss highly asymmetric relationships on the nation-to-nation levels, one could indeed argue that the immediate borderland relationships between the territorial communities on both sides of the international borders is somewhat less oppressively asymmetric. Can transborder regionalism assuage internation asymmetry?

5. Is there any risk that borderland studies may fail to see the forest on account of its focus on only two neighboring trees? I do not think so. We are never oblivious of the fact that two neighbors, facing each other with amity or enmity across the border, have, in turn, neighbors in other directions, even if in some studies on the borderlands in the Southwest we read about the *ambiente*, an integrative border culture, and a special borderland corridor inhabited by 30 million people between Tijuana/San Diego and Matamoros/Brownsville. At one point, Ellwyn R. Stoddard rightly warns, "We depend on one another. To try to separate us will kill both of the Siamese twins."[23] Nevertheless, even such a closely-knit border twosome is embedded in additional concentric circles of influence: the national spheres of Mexico and the United States, the American continent from Alaska to the Tierra del Fuego, and the world system. These successive planes, starting from the border, form an interrelated composite. Thus, Mexico City must concern itself with penetration from not only the northern U.S. border and the southern Guatemala and Belize border, but also beyond them to the rest of Central and South America, and beyond them the world system. In the transborder region along the River Rhine, West Germany interaction with neighboring France and Switzerland is also necessarily affected by the various friendly and unfriendly echoes emanating from East Germany and the Warsaw Pact countries, and beyond them the U.S.-led world system.

As John Donne (1573-1631) expressed it long ago, "No man is an island, entire of itself; every man is a piece of a continent, a part of the main." This concept applies to those millions of people around the world who have to act in and react to the borders — sources of both division and contact. It is inevitable for an island to resent both the interference and the distance of the national and provincial capitals. El Paso and Juárez resent the distance and power of both Washington and Mexico City as well as Austin and Chihuahua City, and their tendency to interfere unduly. "Juárez is an island," states its mayor, Francisco Barrio Terrazas. "There always has been a problem between the border and the (national) government."[24] The local complaint in El Paso is not only that Washington, D.C., is distant and ignorant but that 583 miles separate the city from the seat of its own state government, Austin.

6. Finally, for the purpose of further research, it seems useful analytically to section the borderlands subregionally. The intensity of transborder regional intimacy varies, as Niles Hansen has convincingly argued. Despite the location of the six Mexican and four U.S. states along the border there are many differences in kind and degree; proximity has had varying subregional consequences. In his study of the border economy Hansen distinguished seven subregions of the Southwest borderlands: The San Diego Metropolitan Area, the Imperial Valley, the Arizona Borderlands, the El Paso Region, the Middle Rio Grande Region of Texas, the South Texas Transborder Region, and the Lower Rio Grande Valley.[25]

In the U.S.-Mexican context California is not only affected by the two Baja-Californias but naturally also by Oregon, Washington, and beyond that the Canadian provinces of British Columbia and Alberta. Similarly, Chihuahua and Tamaulipas are not only related across the border to Texas but also to Durango and Vera Cruz, and beyond them to Mexico City. The affairs of border cities are affected by U.S.-Japanese market conflicts and the Central and South American upheavals. As emphasized before, in our interdependent age islands are no longer islands; much less so the borderlands, however isolated and neglected they feel.

Whatever constitutional, conceptual, or practical objections central policy makers may have, noncentralized and decentralized systems have begun to speak with calm but insistent voices across the national borders to foreign central and subnational governments. The issue of adequate coordinative, constitutional, and institutional responses to this polyphony has been placed, and will remain, on the agenda of both practical politics and academic research.

Notes

1. J.A. Caporaso, "Theoretical Framework for the Study of Canadian-American Relations in the 1980s" (University Consortium for Research on North America, Harvard University, 1983).

2. W.F. Hanrieder, "Dissolving International Politics: Reflections on the Nation-States," *American Political Science Review* 72 (1978): 1276-86.

3. S.P. Mumme, "Regional Power in National Diplomacy: The Case of the U.S. Section of the International Boundary and Water Commission," *Publius* 14 (1984): 115-35.

4. National Governors' Association, Committee on International Trade and Foreign Relations, *Export Development and Foreign Investment: The Role of the State and its Linkage to Federal Action* (Washington, D.C.: NGA, 1981).

5. I.D. Duchacek, "Transborder Regionalism and Microdiplomacy: A Comparative Study," unpublished paper (University Consortium for Research on North America, Harvard University, 1983).

6. National Governors' Association, *Export Development and Foreign Investment*, 1.

7. *Perspective* (1983): 2.

8. Duchacek, "Transborder Regionalism," 25.

9. J. Kinkaid, book review, *Publius* 13 (1983): 143-144.

10. Novel research on the "Anglo-French Frontier Region," linked and divided by the Channel, is being undertaken by Richard H. Gibb, School of Geography, Oxford University.

11. S.D. Kransner, "Structural Causes and Regime Consequences: Regimes as Intervening Variables," *International Organization* 36 (1982): 186.

12. J. Steiner, "The Consociational Theory and Beyond," *Comparative Politics* 13 (1981): 339; Arend Lijphast, "Consociational Democracy," *World Politics* 21 (1969): 208-255.

13. F.C. Iklé, *How Nations Negotiate* (New York: Harper and Row, 1964), iii.

14. Francois de Callières, *On the Manner of Negotiating with Princes* (Notre Dame: University of Notre Dame Press, 1963), 56-57.

15. E.R. Stoddard, "Focal and Regional Incongruities in Binational Diplomacy: Policy for the U.S.-Mexico Border," *Policy Perspectives* 2 (1982): 126.

16. J.S. Nye and R.O. Keohane, *Transnational Relations and World Politics* (Cambridge: Harvard University Press, 1972), 39-62.

17. Ibid., 371-98.

18. J.W. House, *Frontier on the Rio Grande: A Political Geography of Development and Social Deprivation* (Oxford: Clarendon Press, 1982), 135.

19. Abelardo L. Valdes, former official of the U.S. Agency for International Development, *New York Times*, 20 March 1981.

20. D. Massicotte, "L'Ontario sur la scène internationale," in *Le Canada dans le monde* (Quebec: Centre québecois de relations internationales, 1982), 90.

21. C. Ricq, *Final Report for the Working Group on Alpine Regions* (Geneva: Insitut des études européennes, 1983), 32.

22. Ibid.

23. *The Border: Special Report* (El Paso: El Paso Herald-Post, 1983), 97

24. Ibid.

25. N. Hansen, *The Border Economy: Regional Development in the Southwest* (Austin: University of Texas Press, 1981), 35-52.

Part II

Western European Borders

———

Border Region Development and Cooperation: Western Europe and the U.S.-Mexico Borderlands in Comparative Perspective

Niles Hansen

Although a large body of scholarly literature exists that maintains that border regions are necessarily economically disadvantaged, evidence from Western Europe and the U.S.-Mexico borderlands suggests that the economic status of border regions can be, and indeed has been, ameliorated by peaceful international political relations and increasing international economic integration. In this context, the interdependent development of neighboring border regions requires local and regional transborder cooperation in order to address effectively social, economic, and environmental problems that spill over international boundaries. The nations of Western Europe have recently agreed formally to promote such cooperation. The United States and Mexico should be able to derive mutual benefits from the lessons of the European experience, because despite obvious differences between the European and U.S.-Mexico border situations, there also are many significant similarities.

Border Regions as Disadvantaged Areas

One of the major themes in the literature on border regions is that they are inherently disadvantaged economically.[1] Fifty years ago, Christaller argued that international political boundaries create artificial barriers to the rational economic organization of potentially complementary areas, and that both the public and private sectors tend to avoid investing in border areas where there is a threat of international conflict.[2] Lösch similarly maintained that political boundaries break up complementary economic areas and reduce opportunities for economies of mass production and free trade. In addition, transportation networks tend to run parallel to boundaries and are competitive rather than complementary. It is particularly noteworthy in the present context that Lösch mapped the financial sphere of influence of El Paso in 1914 (see Fig. 1) and found the same spatial form as he deduced theoretically for ordinary goods, leading him to conclude that "distance affects capital transactions and trade in physical goods in exactly the same way." Finally, he argued that the economies of border regions are distorted by differences in the value systems of political and

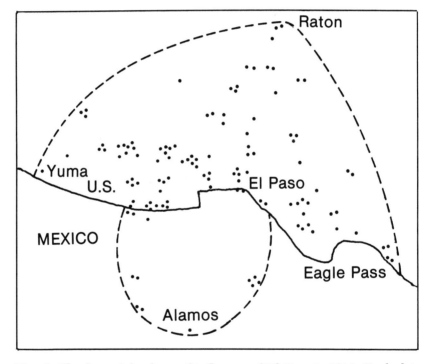

Fig. 1. The financial sphere of influence of El Paso in 1914. Each dot represents one bank keeping an account at an El Paso bank. Source: August Lösch, *The Economics of Location* (New Haven: Yale University Press, 1954), p. 448.

economic orders. The principal goal of the economic order is pros-
perity, followed in rank of priority by *kultur*, power, and continuity.
In contrast, the political goals of the nation-state have exactly the
reverse ranking.[3] French regional-development theorists, addressing
border region issues, also have emphasized transboundary conflicts;
particular attention has been devoted to French border regions that
allegedly are "threatened' by "domination" from neighboring,
economically stronger regions in Germany and Switzerland.[4]

The more recent European literature on border region development
also frequently assumes that such areas are handicapped not only by
border problems as such but also by their extrinsic location with respect
to national capitals, which allegedly neglect relatively distant regions.[5]
Thus, as a recent report of the Council of Europe puts it, "border
regions are the greatest victims of regional imbalances within
Europe."[6] Similar arguments have also been made with respect to the
U.S.-Mexico borderlands.[7]

But is it necessarily true that the economies of border regions are
"naturally" condemned to stagnation? At this point it is appropriate to
examine relevant empirical evidence from Europe.

Border Region Development in Western Europe

Two decades ago, Verburg pointed out that although northwestern
Europe was one of the most prosperous parts of the European Economic Community, there existed within this area

> a belt of insufficiently developed regions marked by
> the fact that they are situated on both sides of national
> frontiers. . . . On both sides of the frontier we find un-
> balanced development caused by the different politico-
> economic systems and by the peripheral character of
> the concerned districts in the national setting. . . .
> The result is that frontier regions remain
> underdeveloped compared with the national centers of
> activity, and at the same time are cut off from their
> homogeneous neighbors.[8]

The border regions defined by Verburg are listed in Tables 1 and 2,
which show, respectively, the growth of gross value added per inhabitant and per employed person between 1970 and 1978 for these regions
and in the countries of which they are a part. Verburg's border regions
included eight of the nine Belgian provinces, the exception being

Brabant, where Brussels is located. In both tables the Belgian border regions are grouped according to whether they are in Flanders (East Flanders, West Flanders, Antwerp, Limburg) or in Wallonie (Hinalut, Namur, Luxembourg, Liège). Data are presented for 1970 and 1978, the earliest and most recent years for which comparable data have been developed. Germany is not included in Table 2 because relevant data were not available at the regional level.

The information in Table 1 indicates that with the exceptions of Groningen and Zeeland in The Netherlands, the *level* of gross value added per inhabitant was lower in each of the border regions than in their respective countries as a whole in both 1970 and 1978. However, in terms of *growth*, the general situation was relatively more favorable. The growth rate of three border regions exceeded the national rate,

Table 1. Gross Value-Added (in European Currency Units) per Inhabitant in Northwest European Border Regions, 1970 and 1978

Area	1970	1978	Percent Change
THE NETHERLANDS	2,190	6,685	205
Groningen	2,588	13,814	434
Friesland	1,673	5,040	201
Drenthe	1,712	5,426	217
Zeeland	2,284	6,821	199
Noord-Brabant	2,061	5,926	188
Limburg	1,814	5,549	206
FRANCE	2,460	6,354	158
Nord-Pas-de-Calais	2,221	5,661	155
Lorraine	2,379	5,805	144
GERMANY	2,909	7,960	174
Weser-Ems	2,265	6,639	193
BELGIUM	2,433	7,085	191
Flanders	2,359	7,039	198
Wallonie	2,109	5,866	178

Source: Eurostat, *Yearbook of Regional Statistics* (Luxembourg: Office of Official Publications of the European Communities, (1981), 117-19.

with that in Groningen being over twice the national rate of 205 percent. In two of the other three Netherlands border regions, the growth rate lagged behind that of the nation by less than 3 percent.

The growth rate of Nord-Pas-de-Calais was only slightly below that of France as a whole, though Lorraine fared less well, while Weser-Ems increased its value added per inhabitant considerably above the corresponding German rate. In Belgium, Wallonie's growth rate was below that of the rest of the nation, but Flanders' growth was more rapid than in the country as a whole.

Table 2 provides an indication of labor productivity. In five of the six border regions of The Netherlands, gross value added per employed person grew more rapidly than it did in the nation. The two French border regions had lower growth rates than France as a whole, but in 1978 the level of gross value added per employed person in Nord-Pas-de-Calais remained higher than the national average. In Belgium, the

Table 2. Gross Value-Added (in European Currency Units) per Employed Person in Northwest European Border Regions, 1970 and 1978

Area	1970	1978	Percent Change
THE NETHERLANDS	6,079	19,906	228
Groningen	7,316	45,219	518
Friesland	5,252	19,279	267
Drenthe	5,425	20,430	277
Zeeland	6,504	22,444	245
Noord-Brabant	6,052	18,785	210
Limburg	5,651	19,011	236
FRANCE	5,990	15,643	161
Nord-Pas-de-Calais	6,276	15,968	154
Lorraine	6,335	15,401	143
BELGIUM	6,353	18,603	193
Flanders	6,338	18,563	193
Wallonie	5,853	16,680	185

Source: Eurostat, *Yearbook of Regional Statistics* (Luxembourg: Office of Official Publications of the European Communities, 1981), 120-22.

growth rate in Wallonie lagged behind the national growth rate, but that in Flanders equalled the national rate.

It should be pointed out that the economic difficulties of Wallonie, Nord-Pas-de-Calais, and Lorraine stem primarily from the fact that these are old industrial regions with heavy concentrations of declining basic sectors, e.g., coal, steel, and textiles. Thus, these difficulties cannot be attributed solely to international boundaries, because throughout Europe and North America many non-border regions with similar industrial bases — including much of the U.S. industrial heartland — have been experiencing similar tendencies.

Another case in point is that of the French region of Alsace, which borders on the German region of Baden, as well as the city of Basel, Switzerland. Although the Alsatian economy has not been as strong as those of Baden and Basel, Alsatians have nonetheless benefited economically from their proximity to these areas. The notion that Alsace is threatened economically or politically by foreign neighbors — an argument frequently found in the French regional development literature of a dozen years ago — is outdated. Indeed, one of the greatest economic threats has been neglect by the central government in Paris, which has been reluctant to have Alsace become more thoroughly integrated into the dynamic Rhine axis, extending from Switzerland to the North Sea.[9] However, the central government's perception of Alsace appears recently to have changed on the basis of the region's relative economic standing within France. In terms of most social and economic indicators, Alsace ranks very high among France's twenty-two planning regions.[10] Ironically, the failure to reinforce Alsace's linkages to the Rhine development axis is now justified on the ground that, in the French context, this relatively prosperous region does not have pressing regional problems.

Taken as a whole, the available evidence does not support the contention that border regions are necessarily disadvantaged, although such may have been the case two decades ago. The barrier function of borders — while very real in the past — has been eroded increasingly by (1) the activities of such integrative communities and groups as the Common Market, Benelux, OECD, NATO, and Council of Europe, and (2) by considerable growth in international exchanges involving trade, worker migration, tourism, and culture — where television has played a significant role. Visas, and often even passports, are unnecessary for travel to a neighboring country.

In Western Europe old conflicts have steadily been submerged in favor of mutual economic development and cooperative efforts to promote the international mobility of people, goods, services, and information, with the result that international boundaries have become increasingly permeable. The importance of transboundary cooperation between border communities and regions has been officially recognized. All of the major governments have signed the European Outline Convention on Transfrontier Cooperation between Territorial Communities or Authorities, wherein they pledged "to promote such cooperation as far as possible and to contribute in this way to the economic and social progress of frontier regions and to the spirit of fellowship which unites the people of Europe."[11]

Border Region Similarities: Western Europe and the U.S.-Mexico Borderlands

While it is still too early to evaluate the practical results of the European Outline Convention, it will be argued here that many similarities exist between the European and U.S.-Mexico contexts with respect to transboundary issues, similarities which could prove instructive, with appropriate modifications, to the U.S.-Mexico borderlands.

The Rhine Basin, for instance, is exemplary for its cross-border and environmental interactions within Western Europe. Because of the need to deal with numerous problems involving international spillovers, cooperative efforts among neighboring border regions have also been especially evident in the Rhine Basin.[12] One of the first — and still the most noteworthy — examples of transboundary cooperation is that of the Regio Basiliensis in the Alsace-Baden-Basel region. The Regio Basiliensis, a trinational coordinating and planning organization, is discussed elsewhere in this volume by Hans Briner, who has been responsible for its remarkable activities since it was established two decades ago. My purpose is to attempt to demonstrate that a basis exists for reasonable comparative perspectives between the Regio Basiliensis area and the U.S.-Mexico borderlands, even though it is acknowledged at the outset that no two border situations are exactly comparable.

The principal difference between the Regio Basiliensis and the U.S.-Mexico borderlands is the degree of disparity in living standards between neighboring countries. In 1980 per capita income was $16,210 in Switzerland, $12,485 in Germany, and $10,709 in France; the corresponding values for the United States and Mexico were $10,408 and

$1,901, respectively. The relatively great difference in the U.S.-Mexico case no doubt contributes to different national and regional perspectives on such issues as international migration and commuting, environmental protection, and economic development strategies. Nonetheless, numerous points of similarity influence transborder interactions in the Regio Basiliensis and the U.S.-Mexico borderlands. This section highlights significant common aspects of admittedly complex phenomena.

1. The U.S.-Mexico border and international borders in the Rhine Basin, including the Regio Basiliensis area, are more a reflection of past conquests and diplomatic arrangements than of natural geophysical barriers.

2. Both border areas were zones of actual or potential military conflict throughout much of the nineteenth and twentieth centuries. As a consequence of the Franco-Prussian War, Alsace, which had been French since the seventeenth century, was ceded to Germany. After the First World War it was returned to France. In 1940 it was again annexed to Germany, only to become French once more with the defeat of the Nazi regime. When Mexico announced it was entering the Second World War, many Mexicans immediately assumed that their country was joining the Germans against the *gringos* because of a long history of conflicts between the United States and Mexico. According to Antonio Haas, "most Mexicans were *germanofilos* during World War II."[13] Since the Second World War, peaceful relations have characterized both borderland areas.

3. In both situations, a Latin national culture (modified by specifically national characteristics) exists beside a national culture that might be termed "northern European." The U.S. culture is frequently termed "Anglo" in the relevant literature.

4. In both situations, international boundaries do not represent a sharp demarcation between border area cultures. The border area is rather a zone where there is mutual cultural interpenetration. In North America this zone has been identified as MexAmerica.[14] Alsace, where many people use a German dialect as their first language, has been accurately described as "profoundly French, but also Rhenish, Germanic, and very European."[15] The following observations concerning the German-Alsace border area have their close counterparts in the U.S.-Mexico borderlands:

   A. In Strasbourg, which lies on the Rhine River, public buses run every thirty minutes across the bridge into the neighboring German town of Kehl, hardly pausing for border formalities at times. (At other times, French customs police stage surprise crackdowns on shoppers who seem intent on buying up cheaper textiles, cameras, or hi-fi equipment in German Stores.)

B. "To us, Bavaria seems more of a foreign country than Strasbourg," a German border policeman has said with some exaggeration, while the manager of the Woolworth's in Kehl claims that the majority of her daily customers are French.[16]

5. Within their national settings, the border regions considered here are all relatively distant from their respective national capitals. It has frequently been argued that officials in the capitals do not understand the unique problems of these border regions and that the central governments, in fact, have helped to create or exacerbate many of the problems found in these areas.[17]

6. In both situations, the Latin national government is highly centralized, though at present the French government is attempting to decentralize more decision-making authority to the regional level. In contrast, the "northern" border regions are parts of relatively decentralized federal states. In consequence, German, and even more so Swiss, border regions have been relatively free to engage in transborder cooperation efforts. The North American setting is more complex. U.S. governors (and mayors) are in fact constitutionally weak in many areas, though "the collective authority of an American state far exceeds that of a Mexican state, which defers to central authority in all substantive things."[18] On the other hand, it has been argued that:

> Mexican border leaders have an advantage over their American counterparts in eliciting government decisions, for they can more easily take their case directly to the top. The decentralized system in the United States requires local U.S. officials to clear several layers of power and a wider network of private-interest groups before they can achieve results.[19]

Recently the U.S. Immigration Service adopted a policy making it more difficult for Mexicans to visit and shop in U.S. border towns. The justification for taking this action without consulting border town officials was: "Why should we consult them? The federal government can't be run by local municipalities."[20]

7. In both situations, the Latin side of the border is not as economically strong as the northern side (although the Latin border area is relatively prosperous within its own national context). This phenomenon has given rise to an extensive literature, particularly in the Latin countries, which maintains that the Latin side of the border is economically handicapped and culturally threatened because of the 'asymmetric interdependence" that exists along the border.[21]

8. Undocumented (illegal) international migration is a significant phenomenon along the U.S.-Mexico border, but it is not an issue in the Rhine Basin generally. However, a considerable international labor force commutes in both areas. For example, in 1982 some sixteen thousand Alsatians commuted on a daily basis to workplaces in Germany and another twenty-one thousand commuted to workplaces in Switzerland. It has been estimated that some fifty thousand "green card" commuters from Mexico work daily on the U.S. side of the border.[22]

9. The argument that the basic force driving transboundary interactions and cooperation is economic development is borne out in the Rhine Basin, as well as along the U.S.-Mexico border. With the exception of San Diego, per capita income in U.S. border metropolitan areas was below the U.S. average in 1981 (see Table 3). However, in five of the seven border metropolitan areas, per capita income *growth* rates between 1976 and 1981 exceeded the corresponding U.S. growth rate of 65 percent. The contiguous El Paso and Las Cruces metropolitan areas were the only exceptions. In terms of population change, all U.S. border SMSA's grew more rapidly during the 1970s than during the 1960s. El Paso, whose 34 percent growth rate between 1960 and 1970 was the lowest for this group, still grew at over three times the national population growth rate of 11 percent. Among all U.S. metropolitan areas, McAllen (56 percent) ranked eleventh in growth during the 1970s; Tucson (51 percent) ranked fifteenth; and Brownsville (49 percent) ranked eighteenth.[23] The phenomenal rate of population growth in Mexico's major border *municipios* has declined in recent years. However, considerable expansion still took place during the 1970s, varying from 26 percent in Matamoros to 35 percent in Tijuana.[24] Moreover, minimum wage rates in Mexico vary geographically, according to level of economic development. In 1983 the highest daily minimum wage rate (523 pesos) applied to all major Mexican border cities.[25] The continuing development of sister cities along the U.S.-Mexico border has generated and will continue to intensify a host of social, economic, and

environmental problems that require transboundary cooperation if they are to be addressed effectively.

10. The presence of a common ethnic group on either side of an international boundary facilitates transborder cooperation. Thus, in Europe it has been noted that

> when the general political trend is towards open boundaries and international integration, the possession of common languages is a great asset for frontier region cooperation. This can be seen in the French-speaking area from Dunkirk to Lorraine, and in the German-speaking area from the Saar to Basel, Constance, and Salzburg; but again it must be stressed that common language does not eliminate problems and conflicts; it only makes their negotiation easier.[26]

In the Regio Basiliensis area, German and Swiss officials frequently can communicate in French, while in Alsace, as already pointed out, German is universally known and even used by many as their first language. Similarly, the clash of Anglo and Mexican cultures in the

Table 3. Per Capita Income in U.S. Standard Metropolitan Areas on the Mexican Border, 1976 and 1981

Area	1976	1981	Percent Change
Brownsville	$3,608	$ 6,172	71
McAllen	3,139	5,606	79
Laredo	3,416	6,148	80
El Paso	4,659	7,360	58
Las Cruces	4,642	7,237	56
Tucson	5,502	9,818	78
San Diego	6,603	10,951	66
U.S.	6,367	10,495	65

Source: U.S. Department of Commerce, Bureau of Economic Analysis, *Local Area Personal Income 1976-81*, vol. 7, Southwest Region (Washington, D.C.: U.S. Government Printing Office, 1983); U.S. Department of Commerce, Bureau of Economic Analysis, *Local Area Personal Income 1976-81*, vol. 9, Far West Region (Washington, D.C.: U.S. Government Printing Office, 1983).

U.S.-Mexico borderlands has been diluted by the fact that most Mexican officials speak English, and even more so because of the large and growing numbers of Mexican-Americans in the U.S. population and among local U.S. officials. The ethnic basis for cooperation is especially strong along the Texas-Mexico border, where informal transborder patterns of contact have been particularly numerous, complex, and friendly in such areas as public safety, social festivities, and public health.[27] The proportion of the total metropolitan population that is Mexican-American in El Paso is 62 percent; 77 percent in Brownsville; 81 percent in McAllen; and 92 percent in Laredo.[28]

11. In both the Regio Basiliensis area and the U.S.-Mexico borderlands, local officials and community leaders have increasingly engaged in informal transborder cooperative efforts. However, these have been largely piecemeal and unsystematic because of the constraints posed by national sovereignty values.[29]

Conclusion

International conflicts tend to inhibit the economic development of border regions. In Western Europe and along the U.S.-Mexico border, however, peaceful international political relations and increasing international economic integration have removed many traditional barriers to border region development. Western European nations have recently committed themselves formally to the promotion of local and regional transborder cooperation in such matters as urban and regional development, transport and communications, energy, nature conservation, environmental protection, education, health, tourism, mutual aid in disaster relief, culture, industrial development, and problems relating to workers who commute across borders. These promising initiatives should provide insights concerning how potential cooperative opportunities can be realized in practice. Despite the relatively great degree of economic disparity that exists in the U.S.-Mexico case, there also are numerous similarities to European border region situations. Thus, in the light of the European experience it would seem reasonable for the United States and Mexico to explore the creation — even if only on an experimental basis — of transborder cooperation agreements that could produce mutual benefits for neighboring border communities or regions.

Notes

1. Niles Hansen, "Border Regions: A Critique of Spatial Theory and a European Case Study," *Annals of Regional Science* 11:1 (March 1977): 1-14; Niles Hansen, *The Border Economy: Regional Development in the Southwest* (Austin: University of Texas Press, 1981).

2. Walter Christaller, *Central Places in Southern Germany* (Englewood Cliffs: Prentice Hall, 1966; original German edition published as *Die zentralen Orte in Sudentschland*, Jena: Gustav Fischer, 1933).

3. August Lösch, *The Economics of Location* (New Haven: Yale University Press, 1954), 199, 448.

4. René Gendarme, "Les problèmes économiques des régions frontières européennes, *Revue économique* 21:6 (November 1970): 889-917; Jacques Boudeville, "Analyse économique des régions frontières," *Economies et Sociétés* 5:3-4 (March-April 1971): 773-90; Jacques Boudeville, *Aménagement du territoire et polarisation* (Paris: Génin, 1972); Sabine Urban, "L'integration économique européenne et l'évolution régionale de part et d'autre du Rhin (Alsace, Bade, Bâle)," *Economies et Sociétés* 5:3-4 (March-April 1971): 603-36.

5. Jacques Robert, *Socio-economic and Physical Planning Problems in the Regions on Both Sides of the Dutch-German and Dutch-Belgian Border* (The Hague: European Research Network for Regional, Urban, and Environmental Planning, 1981); Viktor Freiherr Von Malchus, "Méthodes et pratique de la coopération internationale des régions frontalières européennes," in *Confini e Regioni* (Trieste: Edizioni Lint, 1973).

6. Council of Europe, *Report on Transfrontier Cooperation in Europe*, Report no. CPL (15) 6 (Strasbourg, France: Council of Europe, 1980), 104.

7. Robert R. Nathan Associates, *Industrial and Employment Potential of the United States-Mexico Border* (Washington, D.C.: U.S. Department of Commerce, 1968); Governors of California, Arizona, New Mexico and Texas, *Application for Designation as a Title V Regional Action Planning Commission*, submitted to the U.S. Department of Commerce, 1976.

8. Marinus Verburg, "Location Analysis of the Common Frontier Zones in the European Economic Community," *Papers of the Regional Science Association* 12 (1963): 61-63.

9. Hansen, "Border Regions."

10. Commission of the European Communities, *The Regions of Europe* (Luxembourg: Office of Official Publications of the European Communities, 1981).

11. Council of Europe, *European Outline Convention on Transfrontier Cooperation between Territorial Communities or Authorities*, European Treaty Series no. 106 (Strasbourg, France: Council of Europe, 1982), 2.

12. Raimondo Strassoldo, "Frontier Regions," Background Paper for the Second European Conference of Ministers Responsible for Regional Planning (Strasbourg, France: Council of Europe, 1973); Von Malchus, "Méthodes et pratique."

13. Antonio Haas, *Mexico* (New York: Scala Books, 1982), 20.

14. Joel Garreau, *The Nine Nations of North America* (New York: Avon Books, 1981).

15. Jacques-Louis Delpal, *Alsace* (Paris: Fernand Nathan, 1981), 12.

16. Felix Kessler, "France and Germany, Long-Time Foes, Celebrate New 'Permanent' Friendship," *Wall Street Journal*, 20 January 1983, 26.

17. Hans Briner, *Transfrontier Co-operation in the Upper Rhine Valley* (Strasbourg, France: Council of Europe, Report no. AS/Coll. Front (75)8, 1975); Ellwyn R. Stoddard,

"Local and Regional Incongruities in Bi-national Diplomacy: Policy for the U.S.-Mexico Border," *Policy Perspectives* 2:1 (1982): 111-36.

18. Robert Chatten, *The Conduct of Foreign Relations by State Governments Along the Mexican Border* (Washington, D.C.: U.S. Department of State, Foreign Service Institute, 1983), 35.

19. Oscar J. Martínez, "Mexico's Northern Frontier and the National Political System: Accommodation to Changing Realities," *The Mexican Forum* Special Number (December 1982): 7.

20. *Austin American-Statesman*, 3 September 1983, A-10.

21. Alicia Castellaños Guerrero and Gilberto López y Rivas, "La Influencia Norteamericana en la Cultura de la Frontera Norte de México," in *La Frontera del Norte*, ed. Roque González Salazar (México, D.F.: El Colegio de México, 1981); Urban, "L'integration économique européenne."

22. *Conjoncture alsacienne* (1982): 18; John M. Crewdson, "Border Region is Almost a Country unto Itself, Neither Mexican Nor American," *New York Times*, 14 February 1979, A-22.

23. U.S. Department of Commerce, Bureau of the Census, *Statistical Abstract of the United States, 1981* (Washington, D.C.: U.S. Government Printing Office, 1981).

24. M. Margulis and R. Tuirán, *Nuevos Patrones de Crecimiento Social en la Frontera Norte: La Emigración* (México, D.F.: El Colegio de México, Centro de Estudios Demográficos y de Desarrollo Urbano, 1983).

25. *Mexico Report*, 4:5 (May 1983).

26. Strassoldo, "Frontier Regions," 44.

27. John W. Sloan and Jonathan P. West, "The Role of Informal Policy Making in U.S.-Mexico Border Cities," *Social Science Quarterly* 58:3 (September 1977): 270-82.

28. U.S. Department of Commerce, Bureau of the Census, *Summary Characteristics for Governmental Units and Standard Metropolitan Statistical Areas*, 1980 Census of Population and Housing, PHC 80-45 (Washington, D.C.: U.S. Government Printing Office, 1982).

29. Sloan and West, "The Role of Informal Policy Making"; Strassoldo, "Frontier Regions."

Regional Planning and Transfrontier Cooperation: The Regio Basiliensis

Hans J. Briner

In recent years regional planning has been increasingly influenced by the emergence of development policy making by local decision-making organs of the regions themselves. In this respect regional planning is becoming subordinated to regional politics.

Whether one considers regions within the European Economic Community or in other Western European states, it is clear that the regions are gaining in importance and that they are taking direct control over development mechanisms once directed by national governments or their individual states. This means that a rich variety of Europe's distinctive, self-contained regions will be substantially increasing their influence upon regional development and planning methods. It is useful, therefore, to examine an international border region to show the problems that arise under such regional planning efforts, particularly when the problems cross national borders. This paper deals with the Basel area, a trinational region that overlaps France, Germany, and Switzerland.

Since Switzerland is a small state consisting of cantons with considerable local autonomy, border region interactions across national frontiers have a much greater impact upon it than would be the case were it as large as countries like France or Germany. The smaller a country is, the greater will be the impact upon it of traffic movements, economic transactions, and environmental strains crossing national borders. Switzerland, for example, receives an influx of approximately 100,000 commuting workers daily in such border locations as Geneva, Basel, and Ticino.

It is clear that coordination on all sides of the national boundaries is necessary in order to deal effectively with regional problems; indeed, cooperation and planning across international boundaries is given great attention in Switzerland, which is very active in international organizations such as the Council of Europe. However, since the country is not a member of the European Economic Community, it is presently somewhat handicapped as far as its border relations with Germany, France and Italy are concerned. The example of the Basel Region is therefore especially interesting in terms of the problems that arise in the coordination of planning.

The Regio Basiliensis border region lies in Central Europe between the Vosges and Jura Mountains and the Black Forest. France, Germany and Switzerland each occupy about one-third of the region. Basel has been a transportation center of some significance since the construction of a bridge over the Rhine in 1225. Close cultural and economic ties have existed in this heavily populated area for centuries — long before the existence of France, Germany or Switzerland as nations. The inhabitants speak a common language — Alemannic — a dialect of German. They share a long history as well as common culture, which must now be taken into consideration by appropriately coordinated planning and by the achievement of harmony between the parts of the region. Today Basel is a major European railway and highway junction and a prosperous trade center with an economic base dominated by chemicals, banking, and insurance. This large trinational region of 10,000 km^2 contains over two million people living together, more or less equally divided between the three nations. Table 1 summarizes the distribution of the population between the political jurisdictions involved. Fig. 1 shows the location of the region with respect to the countries of Western Europe, while Fig. 2 shows the trinational configuration of the urbanized area.

The Swiss portion of the region embraces two cantons (Basel-Town and Basel-Country) as well as seven districts of four adjacent cantons (Aargau, Bern, Solothurn, and Jura). The French portion includes the Territory of Belfort as well as the Department of Haut-Rhin; the latter is divided into various *arrondisements*. The German portion includes one urban district — Stadtkreis Freiburg — and two rural districts — Landkreis Lorrach and Landkreis Emmendingen, and the western parts of Landkreis Breisgau-Hochschwarzwald and Landkreis Waldshut. Furthermore, the area includes more than one thousand local governments.

Politically significant differences exist between the degrees of local autonomy enjoyed by these local governments. The French entities are administrative districts of the highly centralized French government; additionally, a department-level "parliament" (Conseil Général) operates but has no legislative powers. The German bodies lie within the State of Baden-Württemburg and are governed under a parliamentary democracy that includes considerable state autonomy. The state legislature (Bundesrat) meets in Stuttgart, the capital. Each of the Swiss cantons has its own parliament with each having significant power to nullify central government decisions made in Bern.

Despite the diversity of governing arrangements and degrees of local autonomy, the various jurisdictions of the trinational region have achieved an unusual degree of cooperation in regional planning. Efforts toward this end began in 1963, with the founding of the Regio Basiliensis, a metropolitan coordinating and planning body involving the participation of industry, three local universities, and the local cantonal governments in the Swiss portion of the region. Two years later a parallel organization was established in Mulhouse for the French portion — the Regio du Haut-Rhin. That same year a three-day International Regio Planners Conférence was held in Basel, at which one thousand planners, politicians, and scientists discussed regional planning models applicable to all of Europe.

From 1971 to 1975, two senior local representatives from each of the three national governments met twice each year to discuss current regional problems. This Conférence Tripartite was replaced in 1976 by the Government Commission for Neighborhood Questions involving a variety of three-party organizations and working groups. Since 1980 the bulk of the work of the regional planning effort has been conducted by specialized working groups dealing with energy, the environment, the economy, transportation, and culture. Fig. 3 shows the relationships currently operating.

Table 1: Population of the Basel Region by Country

	Switzerland	France	Germany	Total
City of Basel	193,000	-	-	193,000
Basel suburbs	182,000	34,000	93,000	309,000
Urbanized Area	375,000	34,000	93,000	502,000
Region Total	580,000	770,000	750,000	2,100,000

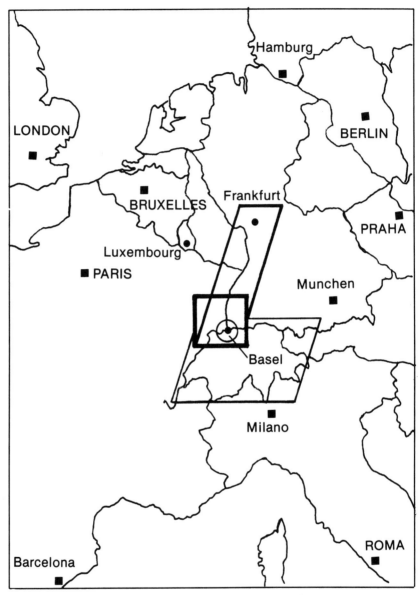

Fig. 1.   The Basel Region, the Upper Rhine Valley, and Switzerland.

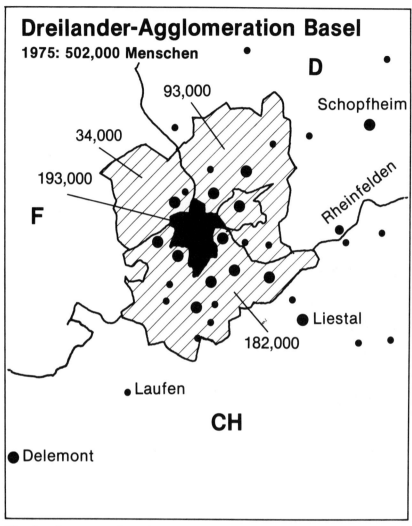

Fig. 2.   Trinational configuration of the Basel Urbanized Area.

The German-French-Swiss Government Commission, the broadest of the bodies created to deal with regional cooperation, was established in November 1975. The Commission, made up of eight members appointed by each of the three foreign ministries involved, develops recommendations for the contracting parties and is able to prepare draft agreements. It has set up two regional committees for the north and south Upper Rhine areas and can delegate the handling of factual matters to various working groups. Questions which interest all three parties are generally dealt with in the southern Three-Party Regional Committee; those which only involve Germany and France are dealt with in the northern Two-Party Regional Committee.

A significant amount of the regional planning effort is achieved at Periodical International Coordination Meetings which take place about seven times per year. These meetings are attended by representatives of the various local planning regions already established in the adjacent portions of the three countries. Another informal group, the Conference of the Upper Rhine Regional Planners, is made up of the fifteen German, French, and Swiss planning associations of the Upper Rhine basin. Since 1972 this body has attempted to outline the basin's development problems and to propose possible solutions. University professors from the three countries are active participants in these conferences, held at least three times each year.

Traffic planning across borders is an important component for the Swiss portion of the region, inasmuch as connections established through Basel have a significant impact upon the overall transportation network of all of Switzerland. Since eighteen thousand workers from France and eight thousand from Germany commute to Basel every day, the coordination of transportation arrangements has been a major concern requiring the attention of the region's planners.

A Swiss commission, under which the three national railways operate, is studying alternatives for long-distance, cross-border rail traffic, including the Basel-Mulhouse rail connection. In view of the newly opened traffic routes in Alsace, which create a three-sided rail route between the Basel-Mulhouse-Freiburg triangle, the communication problems of the region can only be solved through collaboration between Switzerland, France, and Germany. In 1981 the problem of choosing between a Gotthard or Splügen rail tunnel necessitated raising the question about the role of this trinational region in the future European north-south rail system. This consideration creates an additional set of problems; where, for example, in view of the opening of the

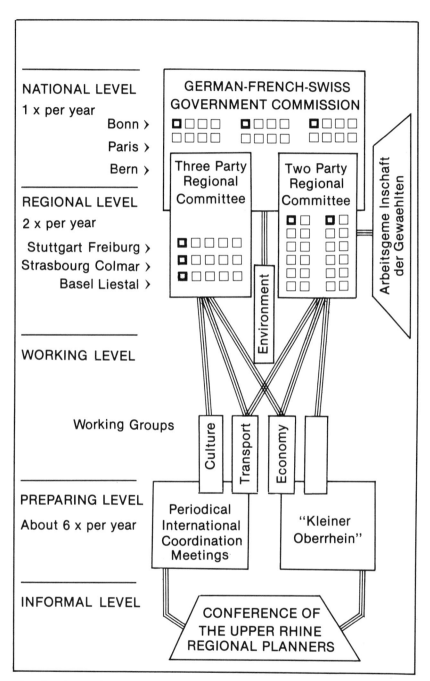

Fig. 3.   Transfrontier cooperation since 1976.

Gotthard road tunnel, can heavy goods can be transferred from highways onto the railways?

The development of the Basel-Mulhouse airport, located in France and jointly operated by Switzerland and France, is another example of a regional transportation concern. Today the airport, used by over one million people annually, involves investments on the order of several hundred million Swiss francs. The residents of all three nations make use of this unique airport.

While considerable success in the coordination of transportation planning has been achieved, questions involving nuclear power and the environment have proven more difficult to resolve. Before any mechanism for regional coordination was created, the combined existing and planned nuclear power plants of the three countries would have resulted in the Basel region having the highest concentration of nuclear power plants in the world, with the use of river water to cool these plants quickly exceeding the biologically permissible cooling capacity of the Rhine. Since the regional bodies lack any real political power, the key to their effectiveness in resolving such sensitive issues lies in their ability to persuade. A six-year study of the environmental impact of the proposed nuclear plants (1975-81) produced a more rational, coordinated approach to the problem that would have been highly unlikely before the creation of the regional planning bodies.

The framework for every internal and cross-border activity of governments, administrative bodies, and private associations, as well as individuals, is set out by the constitutional and administrative system of each country. However, the growing complexity of problems to be solved through "administrative interfaces" today exceeds the scope of the local authorities, leading to an increase in the complexity of administrative duties and to more complex relations between jurisdictions. The economic, social, and infrastructural problems involved not only exceed local authority in the border regions but often exceed national jurisdiction as well. This jurisdictional dilemma requires an interdependence that goes beyond the usual administrative limits. A paradoxical situation arises: the growing requirements of the states lead, on the one hand, to a consolidation of the existing borders at all levels; on the other hand, the economic and social development of the region expands beyond these borders and creates a growing discrepancy between the socio-economic and jurisdictional division of the area.

The study of the emerging relationships in the trinational Regio area reveals several very productive developments in terms of regional cooperation. First, the readiness of the official, politically-established entities to engage in cross-border cooperation and to anticipate pending problems depends decisively on the climate of local relationships at the regional and national level. This climate of cooperation, which tends to be easier to achieve at the regional level, is also constructive at higher levels, depending upon the different internal relationships within each country between the "center" and the "provinces" or regions. Since the border regions can only exercise a limited influence in this respect, international efforts such as the ratification of the European Convention by the Council of Europe are of special importance.

In addition, the creation of a constructive climate of neighborly cooperation is the prerequisite for creating a genuine cross-border cooperation — particularly in the area of regional planning. The existence of such cooperation encourages mutual confidence, but both can be achieved only as part of a long-term effort. Up to the present a genuine, three-sided willingness to undertake a true partnership in the regulation of conflicts in the Basel Region has been missing at the national level — at least in matters such as environmental and energy issues. The initiative for cross-border cooperation therefore tends to come mostly from the regions themselves, and the Trinational Commission (Comission Tripartite) has evidently adapted itself to this perception.

Serious consideration should be given to including members of the three national parliaments on the commissions and other entities established to promote cross-border cooperation. With the exception of the representatives on the Three-Party Regional Committee from the Conseil Général du Haut-Rhin and the Assemblée Nationale on the French side, no members of parliament are represented in the Regio Basiliensis today — a definite handicap.

Without doubt regional planning methods will achieve ever greater importance in the countries involved and in the development of the European Economic Community — and possibly even in East-West exchanges. The coordination of these methods and their incorporation into purposeful European regional policy remains one of the great tasks for the future.

Part III

The U.S.-Mexico Border

Problem Solving Along the U.S.-Mexico Border: A United States View

Ellwyn R. Stoddard

Problem solving is a complex and difficult process even under the best of circumstances. It is further exacerbated when a multinational border region is the setting within which it is undertaken. Both theoretical and methodological dimensions become strained and cluttered as considerations not easily understood by nonborder scholars and jurisdictions arise. Whereas nonborder entities normally focus their analytical skills and resources on substantive solutions to a given problem, border functionaries must expend massive amounts of time and other resources in merely "defining the border." The presence of cultural and language differences, dissimilar national ideologies, and contrasting political or social institutions presents obstacles to transborder articulation. The reaction of external functionaries is often to deal perfunctorily and superficially with these problems, hoping that time and patience will let them slowly disappear. This level of Washingtonesque observation, which defines all national borders as manageable within the confines of universal proscriptions, becomes a polemic for border-based scholars and officials. They, in turn, often adopt a parochial view of border problems by considering them so unique and different as to prevent the application of experience and scientific investigation from other border areas in solving local problems. Recognizing that *every national border is like every other national border* requires the equally important recognition that *every national border is unlike every other national border*. To discover some common ground between these polar viewpoints is the purpose of this essay. The unique historical legacy of the U.S.-Mexico border region will provide the framework from which useful conceptual materials and ideas originating from other border studies may be examined.

From a political-legal perspective, the U.S.-Mexico border is an arbitrary construct separating the sovereign territories of the Republic of Mexico and the United States. This imaginary line delineates a territorial zone of influence which touches at the common border but does not extend beyond it. Each government regards this line as symbolic of its domain within which some evidence of territorial integrity must obtain. As a national institution, the binational border was assigned to those responsible for foreign policy in the nineteenth century and although its function, as well as the world context within which it must be viewed, have altered significantly, nation-states are loath to redefine their policies and procedures with regard to their national borders. Thus, twentieth-century problems must be faced (and presumably solved) within archaic nineteenth-century structures.

Within the United States bureaucracy, operational schizophrenia with regard to the border turns the process of policy formulation into a structural nightmare. While the Departments of State, Treasury, and Agriculture and the attorney general's office monitor the movement of people and products entering this country, they are spurred on by a "fortress mentality," perceiving the border as a barrier which must be maintained between the two countries. At the same time, governmental agencies with responsibilities for increasing foreign trade, tourism, and various facets of international trade seek to make the border a permeable membrane through which people, goods, and services might pass relatively uninhibited. Quite obviously, people representing each of these camps will propose vastly different solutions to the same border "problem"; sometimes seeing a given phenomenon defined as *a problem* by one group expressly makes it *a solution* from the other group's viewpoint. The informal support of illegal Mexican workers within the United States by groups who teach them English, fight for their civil rights, employ them, and recruit their services works against the border control agencies, whereas traditional transnational border intercourse requires the continuation of "informal regulations of border crossers."[1]

Admittedly many other national borders have these same conflicting sturctures and missions to open or close the border to free movement of peoples and things. But the level of formal resistance varies considerably by language and racial composition along our northern and southern borders. Whereas the Canadian-U.S. border is twice as long in miles as the border between Mexico and the United States, along its western expanses, English-speaking Canadians have much less difficulty passing through the border screen than do darker-skinned Mexican Americans through our southern border. Moreover, along the Canadian border more pronounced variations occur in border regulations concerning the French-speaking people of Quebec than among the English-speaking visitors from Alberta. Indian tribes along the Montana-Dakota borderline have much more freedom of movement than do Indian groups along the Arizona-New Mexico-Texas border.[2] Additionally, among the kinship and language groups of newly created African states lying along both sides of the border, and among the Common Market countries of Western Europe (as well as among the Spanish-speaking populations in the American Southwest), formal procedures are designed to separate them at the point where the imaginary national boundary is drawn.[3] But in direct repudiation of these efforts, informal networks remain intact in spite of nationalistic efforts to annihilate them.

Even among borderlands scholars, consensual and divisive elements arise when many disciplines, models, and methodologies are employed for a more complete multi-disciplinary examination of border problems. For example, a geologist, an archaeologist, and a paleontologist might legitimately study the Southwest borderlands, ignoring completely contemporary national boundaries that were quite irrelevant a million or ten thousand years ago. The ethnologist prefers to identify the influence limits of culture areas and the vague frontiers of indigenous peoples rather than to look to modern maps for political-legal demarcations. Even some contemporary historians reject the current national boundary with Mexico as an ideological perspective.[4] Then, in contrast, scholars who deal with current border problems such as smuggling, illegal immigration, border pollution, and similar phenomena are tied to the contemporary binational boundary line as the basis for their investigations. Hence, the term Borderlands (rather than the unidisciplinary Borderland) more precisely illustrates the wide variations in how the current U.S.-Mexico border is conceptualized within the various academic disciplines, by the local and national governments, and within the ethnic groups concentrated along it. Until a common frame of reference can be agreed upon, problem-solving processes will yield few positive results in the management of border problems.

Indigenous and European Conquests
in the New World

A complete chronology of man's development in the American Southwest is not readily available, but enough prehistoric fragments and historical documentation can be surveyed briefly to illustrate some of the adaptive changes required for man's survival in the area. Completely different versions of Southwest history can be derived by careful selection of a given point in time from which to begin one's historical analysis.

A cyclical process of adaptation and conquest[5] began more than ten thousand years ago when Clovis hunters stalked the tapir, bison, and giant mammoth through the swampy lowlands from Arizona to West Texas. Water remaining from Ice Age melts and renewed by monsoon storms began to dissipate, with lush, semi-tropical greenery providing nourishment for these larger animals.[6] Hunters, forced to leave the area in pursuit of the migratory animal herds, left some peoples to adapt to a more sedentary existence in semi-arid surroundings. Pueblo cultures evolved slowly, influenced from the Chalchihuites of Zacatecas/Durango[7] and later from the inmigration of Athapaskan-speaking peoples (Apachean) who began to enter the Southwest from their British Columbia homelands sometime before 1000 A.D. Navaho Apaches lived by raiding the sedentary pueblo villages, ultimately causing them to seek refuge in communities located high on the mesas. Later, tribute to Tewa warriors saved them from further exploitation from without. In a similar manner, the Anasazi fled to southwest Colorado where they built cliff-dwelling communities as protection from raiding Plains and Apache Indians. Thus, prior to the European conquest, patterns of conquest, tribute and warfare had been practiced for millenia among indigenous groups of the Southwest and Borderlands regions.

In Central and Mesoamerica, Olmec, Mayan, Zapotec, Toltec, and Aztec dynasties were put in place and maintained with tribute from conquered peoples. The glories of *Empire*, extolled by our contemporary histories as a sign of cultural greatness, were possible only through the centralization of authority and the economic exploitation of some groups by others whose cultural advances we laud. For nearly three thousand years the more than thirty million inhabitants of Mexico's central valley had served under indigenous lords prior to the arrival of Cortez.[8] Pre-Columbian institutions similar to the *encomienda* and *repartimiento* and to which people had become accustomed facilitated the institution of the European feudal system brought by the Spanish *Conquistadores* to America.[9] New World mining enterprises dictated the relationship of Spanish overlords to indigenous workers. The extraction of gold and silver to fuel Spain's European power was based upon control of land where precious metals might be found. The act of Columbus' landing on the beaches of Hispañola was the foundation for the royal family of Spain to claim the entire Western Hemisphere. In 1514 a formal law called the *Requisito* reclaimed the land as each military commandant informed his proposed Indian prey of his intention to conquer them. He read aloud an edict in Castilian Spanish, explaining how God had placed the Earth in trust of his church leadership, whose representative (the military commandant) now appealed for them to pay appropriate homage. Failure to do so would necessitate taking military action against them because of their rebellion against God.[10] This was an effective philosophical rationalization for the control and extermination of native cultures and the institutionalization of racial social strata.

Peninsular-born whites kept a power monopoly over the Society of Castes, while mixed-bloods were placed in an elaborate hierarchy of categories with full-blood Indians at the bottom and black slaves filling a category beneath the Indians.[11] As the Spanish conquest flowed northward in search of precious metals, the nomadic Indians of semi-arid desert regions and cultures that had never been part of the earlier indigenous Empires did not respond well to Spanish domination. They did not wish to become agriculturalists; therefore, except for the pueblos of northern New Mexico, Spanish control of the land and tribes was limited and generally unsuccessful. Budgetary cutbacks forced Spain to regroup its missions and presidios in isolated regions of Texas and California. The distance from Mexico City also proved to be an administrative problem, inasmuch as the contacts between Mexico City and Chihuahua were rare; trade caravans each year or so to Santa Fe constituted the only continuous contact with that region.

By the beginning of the nineteenth century, the Society of Castes was near collapse, and a general repudiation of domination by Peninsular Spaniards in the New World erupted into revolts throughout Central and Latin America and the American Southwest. Erroneously identified as "independence movements," these revolts were, in reality, a simple substitution of the lords of New Spain for the Old World elites.[12] The social structure changed little.

Today Mexico's social and economic structure continues to reflect the old European system of unequal distribution of income. It is this *internal* exploitation of its wealth and resources by the privileged families at the top which makes American help for "the masses" difficult and uninviting.[13] Even verbal and symbolic claims of Mexico's indigenous origins and pride in its "cultural pluralism" policies are negated by its expulsion of Orientals and callous treatment of its native people in urban centers.[14]

Our United States history of land claims and treatment of indigenous tribes is equally embarrassing when the romantic narratives of current histories are stripped and documented materials are substituted. Early Dutch, German, French, and British intrusions into North America produced a rapidly growing population along the Atlantic seaboard which subsequently proclaimed its independence from European rule. Because the immigrants, from debtors' prisons, sweatshops, or impoverished rural areas, left the feudal vise of European society never to return, they did not have to meet the strict requirements placed on Spanish New World administrators, who were required to have pedigrees showing the absence of Moorish blood for three generations (although for the wealthy there was a way to "purify" their blood line).[15] Anglo immigrants wanted land, free of inhabitants, which they could own and exploit. Therefore, Indian policies were directed at removing Indians from the desirable lands through either relocation or warfare. The courts justified the removal of nomadic Indians through common law, which stated the people could only own land they had worked on; the Indians, as nomads, therefore had no legal right. They were replaced by hard-working farmers who followed the mandate given Adam, "earn thy bread by the sweat of thy brow."[16] Thus, little miscegenation occurred, compared with the *mestizaje* (mixed breeding) found between Spanish lords and their Indian consorts living together in a feudal-type system.

Traditional historical treatment of Indians during this period often overlooks the selective application of democratic principles to the Indian population. Just as democracy, based in Greek philosophy, did not mean that all could participate in the political processes, since that participation rested on a citizenship denied women, slaves, and foreigners,[17] so too were Indians denied the vote. It was not until the mid-1920s that they were allowed to participate in federal elections while living on the reservation. In New Mexico the right to vote was granted only in 1948 for state elections, though slavery had been out-lawed for nearly a century and blacks had gained the franchise, with women also having been elevated to the level of "citizen."

The long history of the Southwest is frequently only peremptorily or inadequately addressed or ignored in standard histories of the United States which are often written as if the nation's story began in New England in the seventeenth century. The Southwest is usually pictured as an area of Mexican bandits and outlaws subdued by brave Anglo-American frontiersmen. The contribution of the Spanish elite of Mexico and California (who led in the fight for independence of those republics), the role played by black soldiers and cowboys in opening up the Southwest, and the part of Orientals and other minorities in the development of the borderlands, until recently were major omissions in virtually every Southwest history textbook. For the traditional power bloc of the Northeast, the U.S.-Mexico border is still regarded as an untamed wilderness where Indians attack white settlers and the "cowboys still walk down the dusty streets of Laredo."

In summary, both Mexican and American histories of the Borderlands are biased to represent the respective ideological views popular with the wealthy classes or powerful political interests within their nations. Each country blames border misfortunes on the presence of the other. The United States is vexed by large numbers of "foreigners" (Mexican Americans) living within its territory and seeks through formal means to keep further immigration from the south under control. Mexicans regard their economic and social woes as resulting from the loss of their claimed lands north of the present border and from economic dependency on the powerful American economic system. Clearly, these perspectives have no common ground, and before any process of problem solving can begin between them, a modified framework, acceptable to both sides, needs to be developed. Historically, America has never felt disposed to compromise since it has always conducted its Latin American policies with unilateral declaration mixed with an occasional "good neighbor" visit. Mexico has never had to explain its internal economic structure in which 5 percent of the population controls more than 40 percent of its economic resources. The policies of the future between these two nations cannot be perpetuated on the myths and antiquated policies of the past. It appears that only a realization by both countries that their destinies are intertwined will change the inadequate understandings, though this proposition does not seem very probable.

A Functional Approach to Border Problems

Often scholars working with border development have sought dynamic models that reflect the processual nature of border interaction, rather than the sterile "international relations" approach that presumes one isolated and autonomous nation lying next to another with only formal treaty ties linking them both. One of these is called the "growth pole theory" and treats economic systems like magnets attracting each other across the national boundary. In the case of the U.S. and Mexico, quite obviously the pull from the more powerful economic system would excessively influence the less powerful one. However, border influences do not always reflect nation-to-nation influences. The highly industrialized center of Mexicali with its smaller U.S. twin city, Calexico, has the "growth pole" going exactly opposite from the influence of the nations on each other. Another model, termed the "integration pole theory," perceives isolated urban centers as cybernetically linked. Pulsations are both emitted and received with constantly shifting reactions from the other centers, in turn causing reactive pulsations to the sender. This model really deals with international economic liaisons, not local ones, and therefore must be modified to be useful for transborder interaction. Stated another way, these macro-theories provide an excellent backdrop against which to evaluate the context to local informal and "formal" cooperative efforts.

To convert these national economic influences into useful border concepts, economists have turned to a neutral trade area orientation which, when operationalized, is termed a "duty free zone." This approach merely separates the single political-legal boundary and its "screening processes" from economic trade. Thus, the original political border's integrity must be maintained, plus *two additional borders* (one edge of the duty free zone located in each nation) must be manned to ensure that the operation of the "duty free zone" does not result in leakage and destabilization within the domestic economy. Sometimes this concept is adopted in practice on a national scale, such as in the Mexican Border Industrial Program (the location of in-bond assembly plants bringing in parts for assembly without duty, shipping the finished product back out without duty) and becomes institutionalized and centralized. Local border jurisdictions in Mexico are responsible for the policies or control of imported materials for the *maquiladoras*, (assembly plants) and the "duty free" zone enhances the *maquiladoras'* supply and delivery channels. This type of national-local coordination of policy and procedures is rare in either country, however.

An organizational approach to border problems encourages the development of binational regional structures through which border problems might be handled and coordination between the two countries achieved. This approach, called "subnational microdiplomacy" or "subnational consociations" by various scholars,[18] is represented along the U.S.-Mexico border by the International Boundary and Water Commission (IBWC). Initially formed by the Treaty of Guadalupe Hidalgo as the commission responsible for final demarcation of the binational border following the 1846 War with Mexico, at the turn of the century it was given the additional task of overseeing the equitable distribution of Rio Grande water and of maintaining the river channels common to both countries. In these engineering duties the IBWC has done remarkably well, but as a model organization for the solution of sensitive border problems, it falls far short. Headed by engineers from both countries, this joint commission refers ticklish political policy problems back to its parent bodies, the Secretaría de Relaciones Exteriores and the Department of State, wherein solutions are then redirected toward national rather than local priorities.[19] Apparently, the mere existence of a joint border structure or binational organization does little for problem solving, unless the essential powers of problem resolution are also included in its mandate and recognized by both nations involved. This recognition, however, involves a value commitment, somewhat independent of organizational structures, procedures, and national posturing.

One functional model relates not only the interaction between two border zones (one each side of the political-legal boundary) but also their relationships with their respective internal state and federal bureaucracies.[20] Enlightening to those engaged in formal operations along the border, it fails to recognize the historical evolution of culture area influences, vague frontiers, and the final emergence of nation-state boundaries and their arbitrary superimposition on existing cultural, economic, and social networks.[21]

A functional border model which claims to be more accurate as an empirical representation of these existing informal and traditional relations has been expressed as a "border buffer zone."[22] Sensitive to the cultural, economic, social, and political interpenetration of nations along their peripheral areas, it postulates a distinct "border culture" with local border jurisdictions from both nations articulating their solutions to common problems with their sister cities or counties through informal (but nationally unapproved) liaisons. It departs radically from the nineteenth century fictions of isolated nation-states, and in place of national self-interest being negotiated through formal treaty arrangements, declares a new doctrine — the *Doctrine of Mutual Necessity* which makes parties *functionally equal* (aside from their relative economic or military power in the world). If border problems are left unresolved, the loss to the stronger nation will be relatively more than to its less developed neighbor. Hence, a symbiotic dependency becomes the basis for border articulation.

Evidence of U.S. and Mexican *interpenetration* is evident to any border observer or tourist. Crossing the boundary into Mexico, one still sees much evidence of trade in dollars, the use of English, advertising of appliances, soft drinks and drive-ins, currently popular rock groups, and other broad manifestations of American culture. Upon returning to the U.S., one encounters Spanish advertising, pricing in pesos, Mexican food, Spanish nomenclature (e.g., rodeo, arena, lasso, etc.), names of towns and cities in border states, and a large Spanish-speaking population using its native tongue.

Because border peoples and institutions deal with immediate problems common to both sides of the border on a day-to-day basis, their reality is geared to *survival*. Arbitrary federal and state proclamations merely treat this system of traditional informal networks spanning the nation-state boundary as "the enemy," rather than the most effective means currently in operation for reducing border tensions and strains. Only a new perspective, such as that found in the Doctrine of Mutual Necessity, can realistically operationalize border problems in such a way as to coordinate national and local jurisdictions in a cooperative ambiance in which effective problem-solving ventures might be instituted and realistic policies formulated. Thus, more critical to the welfare of the Borderlands than nationalistic solutions to its problems is a new and more correct assessment of its *functional reality*.

Coordination Patterns for Subnational Problem Solving

Because border coordination mechanisms are so varied for each region along its expanse, and the problems addressed so different in scope, duration, and intensity, some categorization of these mechanisms or patterns might simplify our investigation. Duchacek's categorization of national-subnational coordination problems along the

Canadian-U.S. border is suitable for examining most efforts at cooperation, although it must be remembered that Mexican treatment by our federal bureaucracy is far different than that accorded to the northern boundary, as previously mentioned. Duchacek's categories[23] are:

1. *High level channels of consultation* between national and subnational structures complementing traditional links found through federalist representation such as senators, congressmen, etc.
2. *Inter-administrative links* comprising formal liaison from federal to subnational governments through which information and coordinative measures are communicated.
3. *Constitutional alterations* or *reinterpretations* that change the milieu within which foreign and domestic policies are conceptualized to correspond more to contemporary world organizations and to subnational efforts to establish formal links with foreign countries.
4. *Direct links through intersovereign organizations* wherein subnational components might approach a specific area of development such as health, education, food production, etc., or a crisis relief situation.

High level consultation. Within the U.S., national-subnational structures such as the National Governors' Association and National Association of State Development Agencies reflect state, not border, interests. On balance, border peoples are poverty-prone[24] and politically impotent. It is unlikely that their miseries, low on state priorities, would be spearheaded in the search for national support for state problems. It is even more unlikely that any national attempts to solve border problems would take serious account of local desires.

When high level diplomatic channels operate *between nations* for the resolution of border problems, it is not the same as national-subnational coordination. For instance, Mexico and United States diplomats met and worked out a resolution to the century-old Chamizal border territorial litigation, and although the treaty signing involved both presidents and was hailed as an unqualified success diplomatically, failure of Washington bureaucrats to have consultation with borderwise officials led to enormous problems for the 1,155 families forced to leave the area to be turned over to Mexico by the 1964 agreement.[25] A similar type of binational consultation concerning border immigration problems in 1978-79 created the "Tortilla Curtain" Immigration and Naturalization Service fence incident which grew out of symbolic national tensions but which had severe borderlands impact.[26] The federal government's concern with the 1982 peso devaluation in Mexico resulted in new legislation as well as hearings on the effectiveness of its "Peso-Pack Program." Yet the wheels of Congress grind slowly. These recommendations have yet to be implemented, even though many of the retail merchants needing assistance have long since closed their doors in backruptcy, as they did after the 1976 peso devaluation, but with more severe and permanent results.[27]

In 1979 Western Europe countries of the Common Market signed an agreement allowing for subnational transborder cooperation and coordination on a formal basis.[28] This pattern has been assessed and recommended for implementation by Nigeria and its neighbors.[29] These two areas, however, should not experience similar success with such transborder structures. Whereas the Common Market experience has driven traditional nationalism to a subordinate position, the dissimilarities between the colonial cultures and experiences of Nigeria and her neighbors would probably make the operation of such structures as difficult as their implementation along the U.S.-Mexico border. Yet even on our southern border there are instances of subnational cooperation for *limited* goals. Regional aviation officials from the U.S. and their Mexican counterparts have entered into an agreement that creates a 30-mile-wide border zone, a neutral airspace, which can be used in cases of bad weather or other emergency conditions by aircraft of either country without prior permission. Only a routine call to the air traffic control center with responsibility for safety conditions in that airspace is required. In a more recent border incident involving low-level radiation in scrap metal brought into the U.S. from Mexico, the Mexican government requested the military at Fort Bliss (El Paso) to fly an army helicopter with detection instruments over Ciudad Juárez to locate any other radioactive metal. Even here, however, a signed document from military commanders was issued before the U.S. Army would dispatch its military aircraft to fly over the neighboring city.

If the high level consultation linkage could be used to set up regional structures with some binational capability for making local *agreements* of limited scope and duration (unlike international treaty-making powers), it would be a most effective step in problem-solving capabilities in border areas. Nevertheless, the mere existence of national-regional linkages, if they do not result in subsequent operations with *quasi-border* autonomy, would degenerate into vehicles for political rhetoric and public display.

Interadministrative links. Such mechanisms have been established in recent decades between the national government and its southern border with only limited success. In 1966 Mexican and U.S. presidents assigned a national representative of ambassadorial stature to their respective border regions. This uneasy formal coordinative mechanism was more a visible display of political good will than a well-conceived and well-financed structure of functional utility. Subsequently, it died from nonperformance and from financial anemia as other governmental requests claimed the scarce public funds needed for all such programs. Federal officials appeared confident that the border would not or could not impact them negatively for their withdrawal of support.

Within the past two decades, regional planning structures affecting more than one state have been allowed by Congress. From this legislation, the Southwest Border Regional Commission was formed to coordinate activities and programs along the counties bordering on Mexico and influenced by it. With the dominating power of California and Texas being "shared" by their sister states of Arizona and New Mexico, this organizational conglomerate of political balance and ethnic appointments was a difficult structure from which to make realistic headway with general border problems. Except for funneling federal money into specific areas of the border, its functions were largely perfunctory and it did not fare well. Had it had a role as a semiautonomous ombudsman, a small body of knowledgeable experts who could interpret border institutions and operations to those less familiar with them, it might have survived.

The ultimate test for such an organization is its functional effectiveness in providing an opportunity for local transnational agreements to be formulated so that local jurisdictions can coordinate their efforts toward resolving their common problems. If the U.S.-Mexico border region had a powerful and popular spokesman, such as Senator Edward Kennedy was for Appalachia, the direct formal liaison structure may have had the support necessary to become effective.[30] But without some federal changes in stereotyping border peoples and border problems within archaic diplomatic and racial-sensitive frameworks, even a formal interadministrative link will not be effective except as a first step to remove federal obstacles to border self-help efforts.

Constitutional reinterpretations.   The world as perceived by our constitutional architects exists no more. No nation is self-sufficient by watching its own boundaries. Territorial borders are no place to safeguard the nation, so through treaties with other nations or groups of nations our extended "self-interest" borders no longer correspond to our geographical periphery.[31] Even during World War II, the U.S. feared surprise Japanese air attacks on the continental United States and sought early warning sites in Baja California,[32] a practice continued with the DEW (Distant Early Warning) line in Canada and listening posts in Turkey and the Middle East.

A century ago no U.S. state carried on direct relations with a foreign power. With the emergence of transnational corporate interests linking raw materials in foreign nations to our own production and competition with foreign nations for markets and factory site locations, this practice has become routine. Recently, then-governor of Texas Bill Clements unabashedly began intrastate and intranational diplomacy with other U.S. and Mexican governors sharing common border problems. The traditional separation of foreign policy from domestic issues has disappeared and transborder relationships often combine the two.

A series of examples illustrates these changes in state-nation relations. The border issues of common air and water pollution problems[33] are enforced as if they were local and ignored when ascribed to inadequate international cooperation. When the Environmental Protection Agency listed the El Paso-Ciudad Juárez basin as failing to meet national standards for clean air, it made no provisions for coordination and cooperation between the two cities. As one national magazine remarked, "It is equivalent to cleaning up the air in two-thirds of the city of Houston."

An informal venture was instituted between El Paso and Ciudad Juárez to locate high pollution areas and to try to curb them through local controls. A truck with highly sophisticated pollution-sensory equipment was taken into Ciudad Juárez each week, the readings utilized by both Ciudad Juárez officials and engineers studying the problem on the U.S. side. While crossing the border one day, this vehicle and its destination became the object of curiosity for inspectors at the International Bridge. Checking further, officials discovered that the act through which federal funds were used to help purchase the truck and equipment specified that it not be used outside the United States. Its deployment in Ciudad Juárez ceased, officially scuttling the informal cooperation toward defining and solving the common pollution problem.

Recently an international marathon run for charity set up a street route leading through El Paso and Ciudad Juárez. A line-painting machine belonging to the City of El Paso was dispatched into Juárez to paint a "blue line" to mark the route, but the painting halted when a legal ruling noted that El Paso and its mayor would be financially liable should an accident occur while in a "foreign country."

An El Paso Juvenile Probation officer was dismissed for, among other things, refusing to obey an order to "unofficially" transport apprehended Mexican national juveniles back to Ciudad Juárez authorities for action regarding lawbreaking offenses committed in El Paso. When she protested her removal for refusing to "go into Mexico," the informal liaison was discontinued by judicial decree.

Though no viable alternative exists to handle errant Mexican juveniles apprehended in the U.S. (state funds cannot be used for their expense in the detainment facility), state and national laws offer no formal solution — only mandates against local informal mechanisms for handling the problem. These incidents demonstrate the archaic nature of legalistic policies currently applied to the border region.

Were constitutional reinterpretations allowed which accept the application of our "coastal buffer zone" concept to land-locked border areas, some headway could be made. The land-border "buffer zone," based on functional interpenetration, might treat "Mexican territory in the U.S." or "U.S. territory in Mexico" as part of our "responsibility boundary" solely for the expenditure of funds and local-agency resource coordination, designed to handle a limited border problem impacting both countries. Then if such issues as a future rabies epidemic arise, the dispatching of American soldiers stationed at Fort Bliss and placed at the disposal of Ciudad Juárez authorities could assist in rounding up the packs of wild dogs (habitating the regions around the *colonias* on the city outskirts), and by controlling their movements, benefit both sides of the border *equally*. Mosquito control, if handled in twin border cities as a single problem, would be more effective and cheaper in fighting "infestation" if only U.S. sprays and machines were utilized on both sides. Clearly, obstructionist constitutional interpretations of foreign and domestic policy must be altered before any meaningful policy-making and problem-solving mechanisms can survive in the border milieu.

Intersovereign organizational links. The establishment of the United Nations produced many educational, health, scientific, and cultural organizations which could function on a multinational basis without excessive diplomatic interference. Although regional organizations such as the Pan American Organization and its health program (PAHO) can provide transborder linkages, their priorities (the more needy Latin American nations) leave them few resources to spend on U.S.-Mexico border health campaigns. A *quasi*-public, professional group, the Border Health Organization, has provided a forum for professionals seeking to learn new techniques, new areas of potential cooperation, and epidemiological information regarding potentially dangerous trends in the future. But the lack of doctor support and participation in such ventures causes the social worker, nurse, and administrator contacts to be of minimal value. Thus it cannot be assumed that by merely establishing a transnational organization within an intersovereign linkage, all local components will effectively use it and gain from its operation.

Through decades of international cooperation during natural disasters, the International Red Cross has been recognized as a symbol of intersovereign linkage. Nevertheless, its acceptance across national borders varies by time and event. In 1955 three devastating tropical storms and hurricanes struck the coastal region of Tampico, Mexico; within a month or so the flooding had destroyed homes, knocked out transportation and communication systems, and made power systems operative only part of the time. The American Red Cross (organizationally an arm of the International Red Cross) combined with the U.S. Navy carrier *Saipan* in providing services and support. In addition to these agencies competing for a major role in the disaster relief operation, however, the lack of a strong Mexican Red Cross organization produced coordinative problems between relief operations and needy victims. When the carrier *Saipan* offered to supply the city with power during a crisis period, local officials insisted upon the official signing of "forms" which caused the naval offer to be withdrawn.[34] The Rio Grande floods of 1954 and 1958 brought forth many of the same transnational coordinative problems seen in Tampico, further complicated by Red Cross and Salvation Army competition to "own the disaster" and by sociocultural barriers between the middle class relief personnel and lower class victims.[35]

In the northern sierras of Mexico, the Tarahumara Indians eke out a bare existence. When a national magazine article described their plight, a group of communities in Louisiana filled some railroad cars with clothing, bedding, canned food, and accessories to go to the needy Tarahumaras. Lacking the legitimation of an intersovereign structure through which to channel the items, they were refused entry by Mexico and eventually the materials deteriorated or suffered from pillage in local warehouses.[36] Only a few truckloads got to their destination by means of "informal networks" interested in Tarahumara welfare.

Findings and Implications

Without some modifications in existing Mexican and U.S. national perceptions of their common borderlands, no organizational mechanism can be advanced which would be effective in helping people of this region solve their common difficulties. Likewise, without some legislative or judicial reinterpretation of "foreign" and "domestic" responsibilities, these changed perceptions cannot be implemented.

A simple organizational solution to border problem solving is somewhat naive. Unless problem solving can function *within the context of local agreements* and cooperation, unfettered by formal diplomatic channels of nation-state bureaucracies, it will be of little use. Moreover, unless federal-level functionaries adopt a *functional view* of border problems and recognize the existing network of transnational informal linkages as a useful ally rather than a legislative target, coordination between local, state and federal levels of government will be superficial and perfunctory at best in border matters.

Both countries place many border issues high on their priority list. These include: Mexican immigration (legal and illegal), peso devaluation causes and impacts, drug smuggling, petroleum and natural gas development and purchases, food and technology exchanges, and others. But unilateral declarations by the U.S., our main weapon of Latin American diplomacy in the past, are no longer effective in Latin America generally and less so in Mexico. Each nation must see that border problem solving can occur only when problems are seen as *a function of the policies and perceptions of both nations.* Unless the Doctrine of Mutual Necessity can bridge the gap created historically by U.S.-Mexico misunderstandings and misinterpretations, by scapegoating, and by insensitivity, no bilateral border mechanism can be developed to allow local border jurisdictions to become intimately involved in coordinating border problems with their local neighbors. Without a mechanism designed for functional utility (not rhetoric and political posturing), local use of this problem-solving system will not be forthcoming. Given the intransigence of border state governments and national officials on both sides of the U.S.-Mexico border, the embracing of a new perspective of their nation and its borderlands does not seem to be very immediate. Short of international crises which would produce the necessity to rethink borderlands policies, a change in the underlying obstacles to border problem solving for our region does not appear to be a very likely prospect, even though small beginnings now might prepare each nation for eventual acceptance of their common needs. They should take a lesson from their respective border peoples who, some time ago, realized their future survival depended upon cooperation, with or without the blessing of their respective nations.

Notes

1. The institutionalized nature of illegal Mexican immigration is discussed in Ellwyn R. Stoddard, "Illegal Mexican Labor in the Borderlands: Institutionalized Support of an Unlawful Practice" *Pacific Sociological Review* 19 (April 1976): 175-210; and Ellwyn R. Stoddard, "A Conceptual Analysis of the 'Alien Invasion': Institutionalized Support of Illegal Mexican Aliens in the U.S.," *International Migration Review* 10 (Summer 1976): 157-89.

2. John A. Price discusses the situation involving Canadian Indians in his *Native Studies: American and Canadian Indians* (New York: McGraw-Hill, 1978). Ivo D. Duchacek comments on the vibrations along the Canada-U.S. border in "Transborder Regionalism and Microdiplomacy: A Comparative Study" (Paper presented to Seminar on Canadian-United States Relations, Harvard University, December 1983).

3. Both Ivo D. Duchacek and A.I. Asiwaju find commonalities in Europe. Duchacek, "Transborder Regionalism and Microdiplomacy"; Asiwaju, "Borderlands as Regions: Lessons of the Transboundary Planning Experience for International Economic Integration Effort in Africa" (Paper presented to International Seminar on the Economic Community of West African States [ECOWAS] and the Lagos Plan of Action, Ibadan, Nigeria, December 1983). Asiwaju also compares the Nigerian borders with the border between the U.S. and Mexico in his *Borderlands Research: A Comparative Perspective* (El Paso: Center for Inter-American and Border Studies, Border Perspectives Paper No. 6, November 1983).

4. For example, Rodolfo F. Acuña, *Occupied America: A History of Chicanos* (New York: Harper and Row, 1981).

5. Edward H. Spicer deals with intergroup hostilities between Amerinds and invading Europeans as "Cycles of Conquest" in his *Cycles of Conquest: The Impact of Spain, Mexico and the U.S. on Indians of the Southwest − 1533-1960* (Tucson: University of Arizona Press, 1962). Frances Leon Swadesh and others examine pre-European inter-Indian conflicts in the Southwest: Swadesh, *Los Primeros Pobladores: Hispanic Americans of the Ute Frontier* (Notre Dame: University of Notre Dame Press, 1974).

6. Lately mammoth remains have been found in Roswell, New Mexico, and West Texas. See the following for descriptions of mammoth and related findings in Arizona: E. W. Haury, Ernst Antevs, and J.F. Lance "Artifacts with Mammoth Remains, Naco, Arizona," *American Antiquity* 19:1 (1953): 1-24; Emil W. Haury, E.B. Sayles, and William W. Wasley, "The Lehner Mammoth Site, Southeast Arizona," *American Antiquity* 25 (July 1959): 2-30, 39-42; John F. Lance, "Faunal Remains from the Lehner Mammoth Site," *American Antiquity* 25 (July 1959): 35-42.

7. Michael S. Foster uses archaeological and linguistic evidence for showing southern influence on the Southwest up to the Great Basin (Utah-Nevada) tribes in "The Mesoamerican Connection: A View from the South" (Paper presented to the Society for American Archaeology, Pittsburgh, April 1983).

8. For an excellent and authoritative view of pre-Spanish Aztec social structure and history, see Miguel León Portilla, "The Concept of the State among the Ancient Aztecs," *Alpha Kappa Deltan* 30 (1960): 7-13.

9. Further examination of these institutions, as well as the role of church missions and military presidios, in stabilizing nomadic Indians as peonage labor pools are discussed

from various perspectives. See Herbert E. Bolton, "The Mission as a Frontier Institution in the Spanish-American Colonies," *American Historical Review* 33 (October 1917): 42-61; Silvio Zavala, "New Viewpoints on the Spanish Colonization of America," in John F. Bannon, ed., *Indian Labor in the Spanish Indies* (Boston: D.C. Heath, 1966), 76-81; Billie Persons, "Secular Life in the San Antonio Missions," *Southwestern Historical Quarterly* 62 (July 1953): 45-62; Lyle N. McAlister, "Social Structure and Social Change in New Spain," *Hispanic American Historical Review* 43 (August 1963): 349-70; Charles Gibson, *Spain in America* (New York: Harper and Row, 1966), Chapter 6; Charles C. Cumberland, *Mexico: The Struggle for Modernity* (New York: Oxford University Press: 1968), 63-64, 83. Also critical is an understanding of Roman law as the institutional legacy of Spain and British Common law as practiced by frontier Anglo peoples. Most of the misunderstandings arising from the Treaty of Guadalupe Hidalgo (1848) were a result of the shift from one legal code to another.

10. Both Gibson, *Spain in America*, 39, and Lesley Byrd Simpson, *The Encomienda in New Spain* (Berkeley: University of California Press, 1950), treat the *Requisito*.

11. See Magus Mörner, *Race Mixture in the History of Latin America* (Boston: Little, Brown, 1967), for an expert analysis of the Society of Castes.

12. Both McAlister, "Social Structure and Social Change in New Spain," and Ralph H. Vigil, "The Lords of New Spain and Mexico," *Rocky Mountain Social Science Journal* 11 (April 1974): 103-12, place this in a realistic perspective.

13. The established Spanish *social* order, based upon racial mixture and the Society of Castes, was altered only slightly during the Revolution of 1910, when it gradually changed to an *economic* order. Today the domination of Mexico's wealth by a few traditional families, the political "sexenio" practice of pilfering the national treasury when leaving the presidency, and ingrained *mordida* are the most serious internal problems with which Mexico must deal. Only "scapegoating" the U.S. focuses public attention away from these internal matters. See Ellwyn R. Stoddard, "Manifest and Latent Consequences of Mexico's Economic Policies − 1982," in *Impact of Peso Devaluations on U.S. Small Business and Adequacy of SBA's Peso Pack Program* (Washington, D.C.: GPO, Congressional Subcommittee hearing, Committee on Government Operations, 9th Congress, 1st Session, 20 May 1983), 379-402.

14. For one case of maltreatment of indigenous women ("Marías"), see Rosalía Solórzano, "Attitudes and Migration Patterns: A Comparative Study of the 'Marías' in Ciudad Juárez, Chihuahua, and Tijuana, Baja California, México" (M.A. thesis, University of Texas at El Paso, 1979).

15. A *Cédula de Gracia* could be purchased by the wealthy in which church authorities verified their "pure blood" even in spite of darker skin color. Gibson, *Spain in America*, 129-30.

16. Vine Deloria, Jr. assembled many regional and federal court decisions concerning Indian land rights and farmer-settler claims. Frequent references to divine mandates and biblical references provided the justification for agrarian interests. Deloria, ed., *Of Utmost Faith* (New York: Bantam Books, 1972).

17. Our popular notions concerning Athenian democracy are quite erroneous. Of the residents of that city-state and its environs, 85 percent were *legally excluded* from democratic participation. Of the remaining one-sixth, half of them were the propertied elite who informally decided governmental policies before they were "debated" in the forum. Warren Breed and Sally M. Seaman, "Indirect Democracy and Social Process in Peridean Athens," *Social Science Quarterly* 52 (December 1971): 631-45; Ellwyn R. Stoddard,

"Worker Alienation — Ideology or Reality? A Critique of Spurious Industrial Worker/Ancient Craftsman Comparisons," revised version of paper presented at the meeting of the Southwestern Sociological Association, Fort Worth (March 1984), Table 1.

18. Ivo D. Duchacek applies the concept of "consociations" to Canada-U.S. border interaction in "Transborder Regionalism and Microdiplomacy." A.I. Asiwaju describes African border problems and variations, recommending "sub micro-diplomacy" for Nigeria and her neighbors in his *Western Yorubaland Under European Rule, 1889-1945* (London: Longman, 1976).

19. For a discussion of the IBWC's handling of water problems vs. policy matters, see Stephen P. Mumme, *Continuity and Change in U.S.-Mexico Land and Water Relations: The Politics of the International Boundary and Water Commission* (Washington, D.C.: The Wilson Center, Occasional Paper no. 77, 1981). C. Richard Bath confirms the agency's technical expertise coupled with its reluctance to take policy initiatives in the area of border pollution matters in his "U.S.-Mexico Experience in Managing Transboundary Air Resources: Problems, Prospects, and Recommendations for the Future," *Natural Resources Journal* 22 (October 1983): 1160.

20. John W. House advocates the "two border zone" concept probably from his political science orientation and implicit acceptance of the political-legal boundary line in "The Frontier Zone: A Conceptual Problem for Policy Makers" *International Political Science Review* 1:4 (1980): 456-77.

21. For discussions of the evolutionary processes of changing culture areas to frontiers, and frontiers to formal boundaries, see Ellwyn R. Stoddard, "Local and Regional Incongruities in Bi-National Diplomacy: Policy for the U.S.-Mexico Border," *Policy Perspectives* 2:1 (1982): 133; Ellwyn R. Stoddard, "Functional Alternatives to Bi-National Border Development Models: The Case of the U.S.-Mexico Border," paper presented to the American Sociological Association, San Francisco (September 1978); Paul Kutsche, "Borders and Frontiers," in Ellwyn R. Stoddard, Richard L. Nostrand and Jonathan P. West, eds., *Borderlands Sourcebook* (Norman: University of Oklahoma Press, 1983), 16-19. Richard L. Nostrand's graphic presentation of shrinking Spanish land claims in North America places the Southwest Session in clearer perspective in "A Changing Culture Region," in Stoddard, et al., *Borderlands Sourcebook*, 6-15.

22. Ellwyn R. Stoddard initiated and operationalized the concept. An elaborate description of "border games" simultaneously played by formal structures and those working within informal networks also utilizes the model. See Stoddard, "Local and Regional Incongruities"; Stoddard, "Functional Alternatives to Bi-National Border Development Models"; Stoddard, *Functional Dimensions of Informal Border Networks* (El Paso: University of Texas at El Paso, Center for Inter-American and Border Studies, Border Perspectives Series, no. 8, January 1984).

23. Duchacek, "Transborder Regionalism and Microdiplomacy," 31-36.

24. For a description of patterns of poverty along the U.S.-Mexico border region see Ellwyn R. Stoddard, *Patterns of Poverty along the U.S.-Mexico Border* (El Paso: University of Texas at El Paso, Center for Inter-American and Border Studies, and Organization of U.S. Border Cities and Counties, 1978). Poverty in the Lower Rio Grande Valley of Texas is examined in Michael V. Miller and Robert Lee Maril, *Poverty in the Lower Rio Grande Valley of Texas: Historical and Contemporary Dimensions* (College Station: Texas A&M University, Agricultural Experiment Station technical report 28-2, 1979).

25. Problems encountered by relocated families as well as Chamizal Act limitations are

discussed in some detail by Ellwyn R. Stoddard, *The Role of Social Factors in the Successful Adjustment of Mexican-American Families to Forced Housing Relocation* (El Paso: City of El Paso, Planning, Research and Development Department, Community Renewal Project, 1970) and "The Adjustment of Mexican American Barrio Families to Forced Housing Relocation," *Social Science Quarterly*, 53 (March 1973): 749-59.

26. The historical setting for this event, as well as contemporary perceptions of local power leaders of El Paso and Ciudad Juárez reveal the incongruencies between federal border policies and local border realities. See Ellwyn R. Stoddard, Oscar J. Martínez, and Miguel A. Martínez Lasso, *El Paso-Ciudad Juárez Relations and the 'Tortilla Curtain': A Study of Local Adaptation to Federal Border Policies* (El Paso: El Paso Council on the Arts and Humanities and the University of Texas at El Paso, 1979). Other glaring examples of federal-local discrepancies in operationalizing border issues dealing with Mexican immigration are the Kennedy-backed Western Hemisphere quota immigration law and the current Simpson-Mazzoli bill, both of which follow the errant federal policy of solving nonlegal border problems with legislation.

27. The impact of the 1976 devaluation on selected border cities is discussed in Ellwyn R. Stoddard and Jonathan P. West, *The Impact of Mexico's Peso Devaluation on Selected U.S. Border Cities* (Tucson: Economic Development Administration, SW Borderlands Consultants, 1977). An overview of the 1982 peso devaluation and its historical context is found in the Congressional hearings on solutions offered by federal agencies. See Stoddard, "Manifest and Latent Consequences of Mexico's Economic Policies — 1982."

28. Some details of this 1979 agreement are given in Duchacek, "Transborder Regionalism and Microdiplomacy," 13.

29. See Asiwaju, "Borderlands as Regions."

30. Although the U.S. region lying next to Mexico includes the five poorest SMSA's found throughout our nation and is unquestionably America's most concentrated "poverty region," national attention has been focused on the West Virginia *Appalachia* as our national symbol for deprivation. Senator Edward Kennedy's staff obtained poverty data during the 1966 subcommittee hearings held in El Paso but chose to champion the unemployed Anglo coal miners nearer to the Northeast. Without a political spokesman, the southern border will continue to suffer poverty without consistent governmental attention.

31. The process of extending national boundaries by means of political treaties is discussed in Harvey Starr and Benjamin A. Most, "The Substance and Study of Borders in International Relations Research," *International Studies Quarterly* 20 (December 1976): 581-620. Currently the U.S. has foreign commitments and "boundaries" in Europe (NATO Pact), in the Mid-East (treaties with Israel and agreements with other countries), and in the Far East (SEATO). Soviet Bloc nations and agreements with Cuba extend Russian boundaries. Black African states exercise similar "political extensions" of their geographic boundaries through treaties and organizational agreements.

32. Mario Ojeda's discussion of this incident brought forth the Mexican sensitivity to transborder cooperation with the U.S. even under potentially threatening conditions — this because the Mexican historical version of mid-nineteenth-century Anglo expansionism on their northern frontiers has been kept alive: Ojeda, *Mexico: The Northern Border as a National Concern* (El Paso: University of Texas at El Paso, Center for Inter-American and Border Studies, Border Perspectives ser. no. 4 (September 1983).

33. Political scientist C. Richard Bath and economist Niles Hansen discuss these border pollution problems from somewhat different perspectives but agree on the ineffectiveness

of federal policies in dealing with them: Bath, "U.S.-Mexico Experience in Managing Transboundary Air Resources"; Hansen, "Transboundary Environmental Issues in the United States-Mexico Borderlands," *Southwestern Review* 2 (Winter 1982): 61-78.

34. Arturo De Hoyos, *The Tampico Disaster*, Unpublished report to the Committee on Disaster Studies, National Research Council (January 1956).

35. Informal relief efforts during the 1954 Rio Grande flood are described by Roy A. Clifford in *Informal Group Actions in the Rio Grande Flood*, A Report to the Committee on Disaster Studies, National Research Council (February 1955), and *The Rio Grande Flood: A Comparative Study of Border Communities in Disaster* (Washington, D.C.: National Academy of Sciences, National Research Council, 1956). For an analysis of disaster relief programs during the 1958 flood of that same river, see Ellwyn R. Stoddard, "Some Latent Consequences of Bureaucratic Efficiency in Disaster Relief" *Human Organization* 28 (Fall 1969): 177-89; and "Catastrophe and Crisis in a Flooded Border Community: An Analytical Approach to Disaster Emergence (Ph.D. diss., Michigan State University, 1961).

36. These donated materials were refused entry at the El Paso-Ciudad Juárez border on instructions from officials higher up. Ostensibly because of possible health hazards or import violations, observers felt that the advance publicity accompanying the project would give the impression that Mexico was unable to care for its own people should the aid be accepted. However, when truckloads of items were quietly taken across the border, they were accepted — but without fanfare.

The U.S.-Mexico Border: Problem Solving in a Transnational Context

Gustavo del Castillo

Introduction

The average individual crossing between Mexico and the United States certainly must ask himself certain questions about the border. Why are traffic lines so long? Are all the inspections and paperwork necessary? Can't the problems of pollution or contaminated rivers on sewage spills be solved in a friendly fashion and in more or less short periods of time? Does it take a major natural disaster such as a tremendous flood or drought, to bring polticians from both countries together to solve such problems?

These are very legitimate questions. They are also complicated ones. The task of the academician is not to confound the problem with complicated, ivory tower explanations but to address these and similar questions in order to derive prescriptive solutions of a realistic nature that appeal to politicians yet do justice to the legitimate (sometimes historical) claims of Mexico and the United States.

The purpose of this essay is not to propose solutions as indicated above, because some of the issues are very complex and the suggested solutions as numerous as the number of interested parties. To propose general solutions would not address the specific nature of most of the issues and would be quite meaningless. Instead, the main purpose of this essay is to signal what are the bases of border conflict and to document the frustrating, serpentine fashion in which decision makers along the border and in Washington or Mexico City have approached issues of public policy and the border.

Most border issues of a transnational nature involve the question of exchange of some kind of human or nonhuman resource where decisions must be made as to the regulation of that resource. For example, flows of people are controlled on both sides of the border; capital, although quite elusive, is also closely monitored. Colorado River water is so managed all along its natural course that it is sometimes impossible to know what has happened to the river — it flows in streams, it is channeled through man-made tubes, it is recycled, etc. The control of the exchange process of resources falls under the purview of formally differentiated structures of governments and of the private sector, but a good deal of this exchange process is managed through informal structures on both sides of the border.

Given this context, this paper will focus primarily on the regulatory actions of formal structures of government, recognizing that these actions probably represent a minor portion of all decisions which are being made with respect to the border. It should also be understood that the articulation sectors and the governments probably represent the more interesting and challenging conceptual problems. In a recent work on the border, the British author John House states:

> At each level there is interaction through the allocation and exercise of governmental powers, certainly also in party political terms and through the interlocking levels of bureaucracies and agencies. Given this complexity and latent dynamism within each country, it is to be expected that there will be sluggishness, contradictions, and tensions when the interactions become international.[1]

Conceptual Framework

There is little question that the U.S.-Mexican border is singular among the many that exist in the world, primarily owing to the dividing line between these two countries separating the largest capitalist nation in the world from a nation struggling for survival. The dividing line is one where poverty meets opulence, and yet in some existential fashion, the border is an ordinary place and crossing it is an everyday occurrence. Beyond the existential moment, however, the social scientist finds considerable conceptual challenges and not a few empirical ones. Real problems exist out there for anyone who chooses to look.

I would like to outline a conceptual framework which will be helpful in looking at border questions. The existence of a dividing line between Mexico and the United States implies immediate questions regarding articulation between the two nations, yet any fruitful conceptualization will be hindered if we visualize the problem solely as one of articulation between nation-states or even between distinct economies. The issue of articulation, which focuses on differential power, becomes crucial if the analysis of border problems is centered around the concept of levels of integration.[2]

If we depart from the premise that multiple levels of integration[3] exist on both the Mexican and U.S. sides of the border, the issue becomes a multi-dimensional one where the border becomes a reality only in terms of the type and degree of articulation existing between a different number of levels of integration. In other words, the units of analysis can be selected from the processes taking place within given levels of integration which articulate with one another, and which the analyst selects as special because of their explanatory power.

In studying U.S.-Mexico border relations we must identify those levels of integration which appear to have the most explanatory power. In this respect the *productive process* is crucial since we can relate it to two of the principal issues between Mexico and the United States: questions of international labor markets and questions regarding the whole phenomenon of migration.

Productive processes occur on both sides of the border but are not independent of one another. They are linked by the flow (the exchange) of human and economic resources. Production is, therefore, a process where transnational integration occurs through the labor force employed, through the utilization of capital, through the transference of technology, etc. Additionally, the productive process is organically tied to the composition, ebbs, and flows of capital.

If we are concerned with the process of production, then a related concern for the behavior of capital along the border arises. For instance, Mexican decision makers had to be concerned with the impact of the devaluations and capital flight of 1982 on the productive process on the Mexican border, not only in terms of whether the product increased or decreased, but also on the multiplier effects which production along the border necessarily has on the Mexican and U.S. economies. It is also clear that the devaluation of the peso has had great effects on U.S. private sector investment in Mexico, which has declined since 1982. On the other hand, if investment generally declined, the

opposite effect occurred along the border where *maquiladoras* sprout-
ed overnight and continued to grow. In this sense one can appreciate
that the border is not only an integral part of Mexico, but that its problems become those of the United States. Therefore, public policy and
the resolution of issues can hardly be the domain of one or the other
country.

Concomitantly, focusing on the social relations of production will,
in all likelihood, give us an insight into explaining that readily observable phenomenon known as "circular migration," identified by many
researchers as that process of international migration of Mexicans who
temporarily leave their place of origin in Mexico to work in the United
States, and after engaging in a process of primitive accumulation of
capital, return to Mexico to carry on their normal patterns of life.
Migration thus becomes a mechanism through which articulation takes
place and where the levels of integration involved in the process can
range from the regional economy of California — say in the Imperial
Valley — or can result in the strawberry fields of Oceanside being integrated with Oaxacan villages or states like Zacatecas and San Luis
Potosí.

Certain important questions arise from such a conceptual perspective. Is it possible that workers find social relations of production so
changed in different work settings between Mexico and the United
States that return to their place of origin is a cultural necessity and
always a welcome event? Is it that they like the Mexican cultural setting or patterns of social relations derived from less advanced capitalist
forms of production? These questions, apparently simple ones, can
only be answered from varying levels of analysis. A straight culturalist
explanation is certainly not substantial enough: that is, an explanation
which alone favors a migrant's propensity for Mexican *fiestas* or hard
drinking makes little or no sense at all. Instead, primary variables such
as economic or agricultural production cycles (associated with rural
religious festivities) are essential to understanding the phenomenon of
migration.

Furthermore, this phenomenon is clearly not independent of public
policies intended to resolve the so-called "problem of the border," insofar as these actions restrict, to varying degrees, the absolute free flow
of populations across the U.S.-Mexico border. In other words, public
policy decisions and actions create the legal (*de jure*) and *de facto*
parameters of the migration issue. In this case, the levels of integration
created by migratory forces are composed of very diverse and complex

actors (or operating units), responding to a myriad of interests and structures. Therefore, if one is interested in deriving an explanation of the process and dynamics of migration between Mexico and the United States, the variables which must be taken into consideration are not only those of "conditions" at the place of origin, or of demand factors in the United States, but also the effects of the formal units of government which encourage or deter migration, such as the cooperation by county sheriffs or city police forces with INS agents. Consequently, one must also ask, what are the impacts on migration of court decisions allowing the education of undocumented children? Or even at a more general level, what impacts on migratory flows does the debate being carried on in the United States over the Simpson-Mazzoli bill have? Does the threat of employer sanctions deter migration? Does an increased H2 program deter undocumented migration in favor of legalizing a worker's stay in the United States? No studies analyze the possible decisions taken by elders in corporate Indian communities as to who will migrate and whose turn it is to stay home. What, for instance, is the importance of the *comisario ejidal* or of local priests in the decisions to migrate within the heavy sending areas?

These are all questions in which semipublic decision makers compose or create differing levels of integration, without which the interrelationships between variables cannot be understood. And while they are also questions of public policy, integration cannot exist without the decisions and actions of semipublic actors.

Another level of integration can be singled out from the interactions between differing systems of power relations and their manifestation at an organizational level. Various levels of analysis deal with political questions along the border that can range from the study of *cacicazgos* in Hidalgo County in Texas's Rio Grande Valley to the apparent inaction of local and state governments faced with Mexican currency devaluations and crumbling border economies. To understand more clearly how our conceptual framework operates in the analysis of a concrete situation, we will now turn to an examination of the 1982 currency devaluation.

A Case Study: The 1982 Mexican Crisis and U.S. Public Policy

Field work has revealed that the most meaningful articulation during the 1982 Mexican crisis took place between local governments and the federal government out of the legitimate demand that the central governments of both nations intervene to provide aid to the affected areas, creating one first level of integration, albeit asymmetrical in nature. Another level of integration took place between the federal levels of government at a binational level to alleviate the impacts of the crisis.

The question here is not why the American federal government actually intervened, but what was the nature of the articulation between different levels of government within one sovereign unit and between the governments of Mexico and the United States.

During the 1982 crisis in Mexico, which greatly affected the border region with the U.S., local governments on the U.S. side saw the impossibility of coping with a crisis of enormous dimensions and sought state aid. Nevertheless, state government offices dealing with questions of regional development were unable to provide even the most minimally adequate solutions, given the dimensions of the crisis which struck border communities in both nations.

The federal government's response in the United States to "the crisis" of 1982 deserves some attention, because it clarifies the nature of the articulation existing between subordinate, peripheral (state and local) units, and the central government. Within months of the first peso devaluation in early 1982, congressional hearings were held all along the border with the aim of determining which sectors had been impacted and to what degree and what kind of aid was required to surmount the crisis. In these hearings a number of actors participated within one level of integration as is demonstrated in the following diagram.

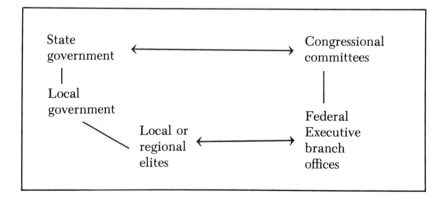

For the most part these hearings led to loan assistance programs for businesses which could afford Small Business Administration loans at 14.5 percent. Apparently few businessmen approached the SBA because of the lengthy paperwork required and because only one commercial bank was lending to affected businesses at lower rates (13.8 to 14 percent). Many of the demands by local elites and businessmen centered around the idea that the border had become a "disaster area" and should receive massive federal relief assistance. Congressmen visiting the border immediately saw a parallel between the border area and depressed zones such as Detroit and Pittsburgh; they countered that if massive relief assistance were to be given to the border, it would also have to be extended to areas like Detroit. This last alternative was clearly impossible to implement and unpalatable to the Reagan administration, ideologically committed to individual progress through individual effort and not through aid packages. Within this context the border would have to be satisfied with limited assistance, coming through normal institutional channels.

Complicating this picture somewhat were the political interests of a Republican administration and its desire to obtain a political clientele from among Hispanic voters (in preparation for the 1984 general election) living along the border, and if possible to help some Republican candidates for the Senate in the primaries being held — especially those in Texas. This clientele would be formed through the announcement of a massive (if fictitious) relief program of border assistance. Before the end of 1982, the Reagan administration had formed the Border Action Group to find a comprehensive solution to the border crisis.

The locus for the resolution of the "border crisis" shifted from the Congress to the Executive branch, within which were two major decision centers. The first, the Office of the Vice-President, was to be responsible for coordinating all aid efforts and programs directed toward the border; the officer responsible to the vice-president was the associate director's Office of Cabinet Affairs. The Commerce Department became the other decision center because it had, according to one informant, a "free, analytic staff which had occasionally been doing jobs for the White House." Within the Commerce Department two different centers were involved in resolving "border problems." One was the Office of Economic Policy, where the studies of the effects of the crisis were conducted; the second center involved the agency charged with implementing the decisions reached in the White House.

Interestingly, the agency charged by the Reagan administration with implementing aid to the border area was the Economic Development Administration, an agency scheduled to disappear under Reagan's austerity program. A Commerce Department source claims:

> This effort (border economic relief) by this administration is fraught with irony. The assistance was to come from EDA, a program which the Administration was committed to closing.

The EDA relief program did not consist of any new funding to be spent along the border but rather consisted of the "regional re-allocation of existing program funds" and the effort to "expedite consideration of programs." Under these conditions, the EDA would offer technical assistance, aid in the use of local planning funds, and help in existing programs like the Emergency Employment Act of 1983 (a public works program). According to EDA administrators, by September of 1983, only some 13 million dollars had been spent to aid the border region.

The reasoning behind the low levels of direct aid to the border had to do with the conceptualization of the real border problem. Those in the Office of Economic Policy perceived the border as an underdeveloped area where solutions have to be long range in order to modify existing structures. A State Department official described the border problem as "a macro problem with micro distortions, before 1982 and after." This generalization was, of course, more relevant to farming areas of the Texas Lower Rio Grande Valley than to the San Diego area.

To respond to the diversity of the border, four regional directors, who were to identify projects and target key areas for treatment, were appointed. They would pass their recommendations on to the associate director's Office of Cabinet Affairs, which was in charge of coordinating the various agencies' programs and see that needs were met by the EDA.

This brief summary of governmental responses to the 1982 Mexican crisis highlights the dominant role of the federal government, some of the political interests represented, and the interrelationships existing between and within the bureaucratic structures in Washington. These interrelationships and the articulations between actors constitute one level of integration leading to an explanation of governmental responses (by the formally differentiated structures) to the 1982 crisis.

In brief, problem solving during the 1982 crisis along the U.S.

southern border fell roundly on the shoulders of the federal government. As the Report of the Southwest Border States Working Group of 6 July 1983 suggests:

> Two principal arguments have been used to justify federal aid to the people, businesses, and governments of the border region. The first is that proximity to Mexico, which benefits the nation at large, has imposed disproportionate responsibilities on U.S. border localities, and that some sharing of these resultant costs is therefore justified. The second argument is that the recent peso devaluations have produced an economic disaster in many parts of the region, overwhelming local self-help capabilities and requiring emergency aid. Both arguments merit scrutiny.[4]

The Role of Power Elites

These adaptive responses by the federal government were certainly not the result of independent initiatives by governmental structures 2,100 miles away from the border. An important triggering element was action by border regional elites.

Why was the behavior of elites important? Because their relationships with the various productive processes, their control of distribution and sale of primary and manufactured goods and services, as well as their control of political offices, made them brokers between the different levels of integration. That is, their actions were not confined to a single arena but touched upon what may be termed strategic areas of regional political economies, effectively linking differing levels of government.

There is no doubt, for instance, that regional political elites in California manipulated the press during the period of the peso devaluation, further aggravating the crisis and reaping untold economic benefits. A participant in the drafting of the final report to the president by the Border Action Group commented:

> It is obvious that a lot of people had made a lot of money
> . . . . they had been raping those communities [cities along the border] for a long time.

Political manipulation also took place, not only in California but also in Texas, resulting in exaggerated political claims by the Reagan administration that aid programs benefited the impacted border areas and that political intermediaries such as Senator John Tower of Texas or Representative Duncan Hunter in California had been key personnel in getting such programs.

This analysis makes clear the existence of at least three levels of integration as shown in the following diagram:

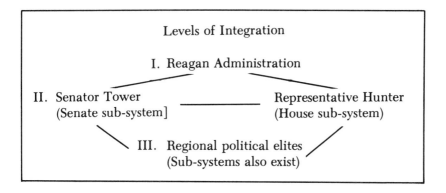

Levels of Integration

I. Reagan Administration

II. Senator Tower
(Senate sub-system]

Representative Hunter
(House sub-system)

III. Regional political elites
(Sub-systems also exist)

One of the principal conclusions to be drawn from the positions adopted by regional businesses and political leaders during the many hearings along the border in 1982 was the unanimous clamor from these groups for protection from ensuing events. The chief demand was that the Mexican federal government adopt some fiscal policies to stabilize the peso with respect to the dollar for the purpose of continuing a steady inflow of dollars in the form of purchases of merchandise. Implied in this demand was the fear that if Mexico did nothing, the American government would find the means to force Mexico to do something: after all, wasn't Mexico responsible for the economic collapse of the U.S. border economies? Weren't the events of 1982 and their manifestations a form of economic aggression against the United States? As always, the American government would find ways to pressure a weak neighbor. To some of these regional elites the border had become a combat zone.

The solutions some of these elite groups demanded were totally unrealistic, indicative of their lack of understanding of the workings of the international economy and of Mexico's needs at a very difficult time. One of these solutions would have had border merchants accepting Mexican pesos and trading them to the U.S. Treasury in exchange for dollars at a stable rate; at this point, the Treasury — now brimming with devalued pesos — would pay for U.S. oil imports from Mexico with pesos.

There is little question that the speculative behavior of California regional elites during the hard times of 1982 not only resulted in economic gains for currency speculators, affecting the short- and long-term value of the Mexican peso, but also resulted in destabilizing the economies of Mexico's northern region.

One common phenomenon along the U.S.-Mexican border during 1982 was the proliferation of currency exchange houses, many of which were not of local origin. In San Ysidro, California, many such establishments, managed by ethnic groups which spoke no Spanish or English, sprang up. Speculation in pesos aggravated the problem of currency flight, as Mexicans who regularly read the U.S. press undertook to secure whatever savings they had in pesos by converting them into dollars. Perhaps even more significantly, this speculative behavior led to a withdrawal of dollars from border Mexican banks under orders from Mexico City thereby giving credence to rumors of further devaluations and leading to further capital flight.

The policies arising in Mexico City were arrived at under a crisis atmosphere and were incorrect ones, given the conditions of the moment. At a time when there was, perhaps, a legitimate demand for dollars, the banks in the border region should have made dollars available to buyers in order to quell devaluation rumors and to show that policies were being implemented to stabilize economic conditions in Mexico. What is clear in this example is that at that time (September-October 1983), processes occurring at the border and events in Mexico City and Washington were increasingly integrated.

No doubt a complementary desire in Mexico City to stabilize the value of the peso existed. However, this desire had little to do with the interests of U.S. border merchants to continue making a profit from sales to Mexicans. A stable currency was in the Mexican national interest, and it certainly also aided international commerce, especially that of the border area. The actions of currency speculators ran contrary to Mexican national interests and did not help border commerce. Still, border elites profited from both approaches, asking for aid in protection of their business interests, as well as speculating in pesos.

The purpose here is not to make normative judgments over the actions of unscrupulous Mexican or American businessmen, yet the actions of elites, with respect to the peso devaluation in 1982, demonstrate a level of integration which surpasses the local or regional level.

The "necessity" to stabilize the peso in relation to an increasingly overvalued dollar (in terms of world markets as well as the result of the massive peso devaluations) brought about the integration of major world actors, and because of the asymmetrical articulation between Mexico and these actors, major changes in Mexican fiscal and monetary policy took place.

The least important demand in this respect came from border regional elites for a stable peso; yet these demands were closely needed by local, regional, and national representatives, who made their concerns known in Washington, which, in turn, pressured Mexico on the imperative that the peso value be stabilized. In order for this stabilization to occur, Mexico had to seek international aid from the U.S. Treasury and the International Monetary Fund.

To obtain aid, Mexico made and has continued to make major changes in its fiscal policies. The return to relative normalcy has again affected the life of border residents. Monetary controls are gone now (after having been imposed for a short period of time by a desperate López Portillo administration). Pesos by the millions are being exchanged for dollars as international commerce begins to develop anew.

Actions by world and national actors have had regional impacts. During the crisis of 1982, the constitution of differing levels of integration, involving regional, national, and world actors, became manifest. While these levels might have been articulated for only a short period of time, if we choose to disregard this process of articulation and pretend that events at the border were the result only of a devaluated currency, then our understanding of the complexity of life at the rim, at the U.S.-Mexico border will be incomplete.

The argument throughout this paper has been that a formal approach focusing on the policies and actions only of formally differentiated structures during the crisis of 1982 overlooks the behavior of actors operating in political and economic arenas not covered by formal analysis. The more formal approaches disregard (because their analytic tools do not permit them) those social dimensions where cultural expressions take place or where political and economic actions are not differentiated. In other words, even the economic analysis of time series data per se does not demonstrate processes. An analysis of variance can only give results of discrete differences but no explanation behind the variance. This paper has introduced two theoretical concepts which can give a richer understanding of border phenomena. Nevertheless, the utilization of these concepts requires methodologies involving primary field research which has not been done in the border region, with some clear exceptions.[5]

The analytical constructs of levels of integration and articulation refer specifically to processes and do not necessarily focus on the particular behavior of formally differentiated structures, although most analyses of public policy and decision making tend to do so. For example, to focus solely on the Small Business Administration or the EDA, when seeking explanations of the problems of economic crisis along the U.S.-Mexico border, appears unduly restrictive and formalistic.

The more formal approaches clearly ignore those structural elements which have made the border area susceptible to Mexico's economic dislocations. Instead, the variables which we must examine necessarily involve the element of process: productive endeavors, social relations of production, conflict and accord among elites.

Problem Solving at the U.S.-Mexico Border?

Truly bilateral efforts have not been developed to deal with the most recent economic dislocations along the border. This is because Washington officials have conceptualized the events as problems of regional development, not differing from those of Detroit, Appalachia, or other economically depressed ares of the United States requiring unilateral solutions from a sovereign government. In effect, for these officials the border is a latent entity, a distinct line separating life styles and economies. This conceptualization persists even when the existing integration and interdependence of U.S. and Mexican border is acknowledged.

If border problems are indeed of a regional nature, then problem solving is, of necessity, a bilateral matter; regions do not recognize artificially drawn boundaries. This fact of life makes problem solving difficult for local authorities, limited by information gaps about the economic behavior of contiguous government units and by limited staffs. Yet it is important to clarify through careful empirical research (usually out of the boundaries of local or regional units) whether phenomena such as the 1982 Mexican crisis were, in fact, of regional proportions or whether the effects were localized and contained.

Research carried out by this author under the auspices of the *Centro de Estudios Fronterizos del Norte de México* (now Colegio de la

Frontera) on this particular topic indicates that in California, cities located twenty miles north of the border were not affected by the 1982 Mexican currency devaluations. Rather, retail sales actually increased during the period of currency controls (the fourth quarter of 1982). More important for this discussion, data about the regions of the Imperial Valley and San Diego gave little or no indication of the actual economic behavior within cities in these regions. To comprehend the phenomena of the impacts of the 1982 devaluations on the border region, data had to be disaggregated into units such as cities and specific retail items. In other words, the region as a unit of analysis was inadequate to explain the diverse nature of what was occurring along the border.

Interviews with local officials in San Diego, Chula Vista, National City, and cities in the Imperial Valley indicated that they assumed that retail sales in their area would plummet, with the corresponding sales taxes decreasing, consequently lowering the income base with which they provided local services. Some of these officials were prepared to cut school district budgets and terminate the employment of school and other public service personnel because of a supposed shortage of funds — all of this on an a priori assumption about the impact which Mexican shoppers had on local economies.

Returning to the question of problem solving, one should ask a number of questions about the operations of government at all different levels. An apparently simple question is why Washington decides to become involved in some issue areas and not in others. This question raises a profound issue about the limits of government intervention, especially along the border area, where problem solving takes on transnational characteristics. If limits do exist, what happens when specific government units tackle issue areas they are unprepared to resolve? What are the consequences for conflict or cooperation along the border? For instance, during the 1982 crisis, was Washington's conceptualization of the problem at all correct? What was Washington's understanding of the devaluation of the peso as it manifested itself along the border? Was there variation between what was happening in Brownsville and San Ysidro? Real understanding of border phenomena has often been confused or substituted by unproven assumptions on the part of border residents, the press, and academics who, for example, have incorrectly assumed in an a priori fashion, that the U.S. border economy is in effect an appendage of the Mexican economy.

The argument which should be presented is that problem solving cannot occur when the parameters for decision making are based on ignorance. That ignorance is not limited to Washington. Border zone officials, at least in California, are limited in terms of the information they have and in terms of the conceptual systems of border phenomena within which they operate. Given this context, they prefer to seek what appears to them as the necessary expertise from federal agencies and services instead of seeking local help. Federal help is also sought with the knowledge that Washington can be much more effective in dealing with Mexico City: at least in 1982, the implicit belief was that a correct option under the circumstances was U.S. pressure toward Mexico.

This situation does not change dramatically when the private sector continually seeks the assistance of Washington representatives and lobbyists. Washington acts in effect as a broker between the U.S. border region and Mexico. Because of the centralized character of Mexican federalism, Mexican border entities must rely on Mexico City, even though they may prefer to achieve a certain degree of independence from the center.

What becomes clear from this discussion is that problem solving along the border is becoming increasingly difficult. An example of the complexity of border issues can be seen in the controversy between the cities of San Diego and Tijuana. A conflict has arisen over the question of Tijuana's waste disposal, which contaminates San Diego beaches. The waste obviously originates in Tijuana, but it is beyond the competence of municipal authorities there to resolve the problem. Nevertheless, San Diego government officials continue to seek their participation in negotiations directed toward resolving the problem. Unfortunately for the two communities, the levels of integration which develop in this case are not between Tijuana and San Diego, but between Tijuana and Mexico City and between Mexico City and San Diego.

Problem solving and the development of public policy are becoming more complex because of the tendency at the local and federal levels to articulate and constitute single levels of integration, and perhaps even worse, the apparent willingness to link issue areas. In the case of Tijuana waste, local and federal officials have suggested that a trade embargo be imposed — not on Tijuana, but on Mexico — until Tijuana resolves the waste problem. This demonstrates that local officials in California fail to understand the upper limits of Mexican border units, while employing federal limits to regulate U.S.-Mexico border relations.

Notes

1. John W. House, *Frontiers on the Rio Grande: A Political Geography of Development and Social Deprivation* (Oxford: Clarendon Press, 1982), 239.

2. The concepts of integration and articulation have been utilized by anthropologists for a good number of years now; they derive from two distinct theoretical traditions. The concept of integration was utilized early by British anthropologists to describe the unity of primitive African societies. Perhaps the most complex treatment of the concept of integration was made by Steward, who goes well beyond the notion of a harmonious working of elements in a society. Steward incorporates productive and political processes as key explanatory variables in the evolution of societies. Once the recognition had been made of the importance of these two elements, the concept of articulation was accepted and utilized as a derivative of integration, in those situations where there are power relationships; in this sense this concept arises out of late Marxist theory. Rosa Luxemburg was one of the first to use the term in this fashion where she proposed the articulation between central-core capitalist nations and their colonies: articulation is utilized not as a substitute for "integration" or of "interrelationship" but as a category indicative of power differentials. In this sense, articulation refers to the interrelationships existing between two levels of sociocultural integration. This is the sense in which this category will be used throughout this paper. This use of the concept also varies from the way Adams utilizes it. Julian H. Steward, *The Theory of Change* (Chicago: University of Illinois Press, 1973); Richard N. Adams, *Crucifixion by Power* (Austin: University of Texas Press, 1970).

3. Steward, *The Theory of Change.*

4. Southwest Border Working Group, *Report of The Southwest Working Group* (Manuscript presented to the president of the United States, 6 July 1983), 69.

5. See William D'Antonio and William H. Form, *Influentials in Two Border Cities* (Notre Dame: University of Notre Dame Press, 1965).

Trans-Boundary Ecosystem Management in the San Diego-Tijuana Region

Lawrence A. Herzog

It is now widely recognized that the United States-Mexico boundary zone is the most urbanized border region in the world. More than seven million inhabitants live in cities on the international border, while a similar number resides within a sphere of influence of one hundred miles on either side of the boundary. The border region is no longer a "frontier," or area of transition between one nation-state and another; it is a territory of dynamic economic establishments, regional urban centers, autonomous subcultures, and emerging national recognition.

The appearance of major cities on the landscape of a region once termed the "land of sunshine, adobe and silence," has injected a vital new agenda into United States-Mexico foreign policy negotiations, the agenda of border policy issues. One need only look at the minutes of recent Mexico-United States Interparliamentary Conferences, held annually, to grasp the degree to which border policy issues such as groundwater rights, air and water pollution, sewage management, tourism, and trade have become elevated to high priority concerns in the field of U.S.-Mexican relations.[1]

It is no accident of geography that a distinct symmetry characterizes the pattern of urbanization along the two-thousand-mile United States-Mexican boundary. The pattern consists of sets of paired urban centers on either side of the international boundary at strategic locations. The term "twin cities" has been applied to these bicultural metropolitan areas,[2] although emphasis should be placed on the inter-connectedness of the settlements, rather than the state of being identical. The important characteristic of these paired urban places is that, because of a common heritage and history, perhaps stated best in Martínez's detailed history of Ciudad Juárez,[3] they have evolved in an interdependent manner.

Although separated by an international boundary, the paired cities of the United States-Mexico border are socially, economically, and functionally integrated. We might speak of these paired cities as "ecosystems," in that they represent human habitats in which many of the biological and physical components (hydrological systems, air sheds, land formations) and many social processes (consumption, family structure) behave in a manner that both transcends and ignores the international political boundary. In the words of a French scholar, "All boundaries are by their nature artificial and can only be viewed as an invention of the human mind. Lines may be a topographical convenience, they are not natural facts. Nature abhors lines."[4]

Thus we must recognize that twin city "ecosystems" are the most important byproduct of the twentieth century urbanization of the United States-Mexico borderlands. In drawing the boundary between the two nations in the nineteenth century, however, before significant settlements sprang up on the regional landscape, both nations understood the fact that the border region was a delicate ecological setting requiring careful management. One of the most important treaties between the two nations, signed in 1884, attempted to outline a framework for maintaining a boundary line in the Rio Grande and Colorado rivers, given changes which the forces of nature might impose on these physical geographic structures. Later, in 1944, a second major U.S.-Mexico treaty addressed the ecological concerns affecting the utilization of the three major border area rivers, the Colorado, the Rio Grande, and the Tijuana River.[5]

Recognition of twin city regions as "ecosystems" is the first step toward understanding the prospects for management of the growing set of problems affecting twin city regions on the U.S.-Mexico border. The notion of an ecosystem reflects the inherent transboundary linkages that require a bilateral management effort. These linkages suggest the kind of "international symbiosis" that Price spoke of in writing about Tijuana, Mexico, over ten years ago.[6] This essay offers a conceptual model of the U.S.-Mexico Transboundary Urban Ecosystem, considers empirical evidence of this ecosystem for the San Diego-Tijuana twin city region, and outlines some considerations vital to developing a transboundary management framework for the U.S.-Mexico urban ecosystems.

A Conceptual Model of the U.S.-Mexico Transboundary Urban Ecosystem

Fig. 1 offers a conceptual model for border city ecosystems. The ecology of the border region is conceptualized on three distinct levels: the natural environment, the built environment, and the human environment. The juxtaposition of these three levels reflects the essential "man-land" relationships characteristic of all geographic regions. The built environment in this model portrays the material imprint of the relationship between "man" (human environment) and the "land" (natural environment) evolving over time. The model seeks to account for the entire spectrum of sociobehavioral and environmental components of an ecosystem.

Most crucial to this model, however, is the fact that the ecosystem spills across a man-made political boundary, the international border. Therefore, at each level of the ecosystem (natural, built and human environments), transboundary spillover effects, or "externalities," must be accounted for as outputs of the ecosystem requiring management. Examples of U.S.-Mexico transboundary externalities are provided in Fig. 1.

Externalities reflect a principle commonly utilized in public choice theory and microeconomics to describe a situation in which an event occurring in one political jurisdiction spills across that jurisdiction's boundaries into a neighboring unit of government.[7] From a public policy perspective, negative externalities require local governments to regulate the problem by assessing a penalty, such as a tax or a regulatory procedure, upon the private firm responsible for the externality. For example, a steel factory causing pollution of a nearby river might be required to pay a special penalty fee to local government for cleaning up the polluted river, or the factory might be required to build new pollution control facilities. In either case, it is the local jurisdiction that imposes this regulatory measure on the source of the externality.

The unique character of the transboundary ecosystem results from the fact that externalities spill across the political boundary, making it

Fig. 1
TRANSBOUNDARY URBAN ECOSYSTEM MODEL

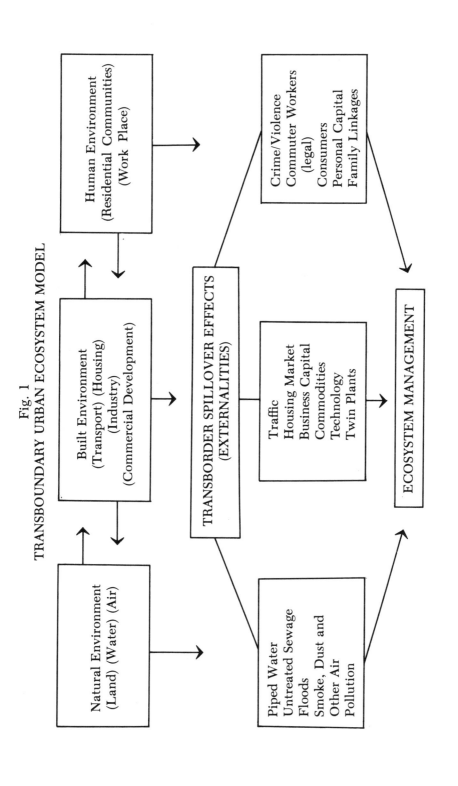

more complicated for political jurisdictions to regulate negative externalities. Within a nation or region, communities or political jurisdictions are usually able to agree on a policy solution that "internalizes" the negative externalities. In some cases local jurisdictions will band together to form a regional administrative entity to deal with externalities that have region-wide influence. Air pollution districts, regional sewage authorities, and regional transportation planning agencies are a few examples. But in an international border region, the problem of externalities is complicated by the differences between the political cultures of neighboring nation-states and the complexity of achieving international coordination of externality effects.[8]

In this way the dilemma of managing transfrontier externalities becomes the central feature of the conceptual border ecosystem model. Where the management of externalities between two local jurisdictions is generally a simple public policy task, management of externalities on an international political boundary puts decision makers into the domain of international relations and foreign policy. It now seems clear that many urban planning questions along the international frontier between the United States and Mexico will, in the future, become foreign policy matters as well.

The border ecosystem model in Fig. 1 recognizes that on all three levels — natural, built, and human environments — neighboring U.S. and Mexican border cities are increasingly situated within a reciprocal sociogeographic milieu of activities. At the level of the natural environment, such topographic phenomena as watersheds, mountains, canyons, and air basins are superimposed over the political boundary. It is now clear that such twin city regions as El Paso-Ciudad Juárez, Brownsville-Matamoros and San Diego-Tijuana share the same air basin, making the control of air pollution an international issue. Twin cities often share a common watershed, since natural hydrological systems were formed long before the Treaty of Guadalupe Hidalgo defined the international boundary in 1848. Such important rivers as the Rio Grande, the Colorado, and the Tijuana are all shared by U.S. and Mexican settlements along the border, while many land formations transcend the boundary: mountain chains, flood plains, canyons, mesas, and river valleys. Many important planning decisions affecting the control of flood waters, public facility location, or land development are made in areas where a land formation affects a settlement across the border. Land use decisions become international decisions and urban planning policies are elevated to foreign policy matters.

At the level of the human or social environment there is considerable evidence of growing symbiotic relations between settlements north and south of the U.S.-Mexico border. The border cities represent, in many ways, the most vivid testimony to the historic and cultural ties between Northern Mexico and the U.S. Southwest. Strong family networks traverse the political boundary. Mexican culture, spilling north into U.S. border cities, is visible in that city culture, if one looks at such evidence as language and architecture. The influence of U.S. culture in Mexican border cities is equally pervasive and shows up in the architecture, commercial activities, language, and technology of those settlements. The most visible sign of an international social environment in the border cities is the human movement that transcends the boundary. Because of economic incentives, Mexicans regularly cross the boundary to seek employment in the U.S., while American consumers seek products and services that are less costly in Mexico. Additionally, business relationships are increasingly bilateral. In short, the burgeoning volume of transboundary movement in such places as El Paso-Ciudad Juárez, Caléxico-Mexicali and San Diego-Tijuana attests to a gradual evolution of truly international border city social areas.

The built environment lies at the intersection of the human and natural environments, since it is the product of man's impact on the natural setting, the surface upon which man-made cities are built. Harvey refers to the built environment as the "totality of physical structures — houses, roads, factories, offices, sewage systems . . . a mass of humanly constructed physical resources broadly appropriate to the purpose of production and reproduction."[9] Along the United States-Mexican border, the built environment of cities provides the most convincing indicator of the degree to which cities have become interconnected. When the physical plant itself in border cities becomes a bicultural entity, then one can argue that the natural and social environments of the border region ecosystem have become so international that man-made, political acts are taken to construct a binational infrastructure. Thus we find emerging evidence along the border of a built environment that is bicultural.

Empirical Components of the San Diego-Tijuana Ecosystem

In order to examine more systematically empirical evidence of the growing symbiotic character of the San Diego-Tijuana ecosystem and its evolution over time, an archival study of newspaper reports covering the San Digeo-Tijuana region from the period 1950 to 1984 was undertaken. Using newspaper stories on border related topics in the two major newspapers of the region, the *San Diego Union Tribune* and the *Los Angeles Times*, articles written during this period were collected through indexing systems, and classified within the three environmental categories of the ecosystem model (natural, built, human). More than two hundred different newspaper reports were used in the archival search.

A comment on the uses of this sort of archival data should be offered. Newspaper reports do not provide an entirely objective, standardized historical account of events having a uniquely transboundary impact. Likewise reports in the San Digeo-Tijuana region did not always reflect precise coverage of important local events. Reporting and publication of information was subject to changing editorial policies within newspaper organizations regarding coverage of U.S.-Mexico border issues. Equally, reportage of events may have reflected changing resources within newspapers at various points in time. All of this suggests that our analysis of changing patterns of border spillover effects in San Diego-Tijuana was subject to certain intervening forces.

Nevertheless, newspaper reports provide one of the few consistent sources of documentation of border issues for the San Diego-Tijuana region and were therefore deemed useful to the investigation. Time and resource constraints and the lack of available archival indices prevented the use of reports from Mexican newspapers in the archival study.

Tables 1 and 2 summarize the results of the archival search for newspaper reports dealing with the San Diego-Tijuana region. In general, one sees that ties generated in the natural environment have caused local concern for the longest time. These ties essentially are derived from the Tijuana River watershed, which originates in the U.S., crosses into the mountains east of Tijuana, passes directly through the center of Tijuana, and drains across the boundary into the U.S., and out into the Pacific Ocean. This watershed has generated periodic floods and sewage spillage problems dating back to the 1950s and earlier. A second set of newspaper reports, dating back to the 1960s, mentions the problems of narcotics smuggling and border crime.

Tables 1 and 2 also suggest that historically, the spillover problems affecting the San Diego-Tijuana region were related to either the physical environment or the social environment. It was only in the

Table 1.   Dominant Border Planning Issues Reported (By Year) San Diego-Tijuana

to 1958	1960	1965	1970	1976	1980-1984
Flooding Sewage	Sewage Floods Narcotics/ Smuggling	Sewage Flooding Tijuana River development	Urban Planning and growth Tijuana/ San Diego	Peso devaluation Illegal immigration Channelization of Tijuana River Baja coastal development Border crime Free Trade Zone/Business Land Use: Otay Mesa/Mesa de Otay Tijuana River Channeliza- tion/displace- ment of residents	Crime and violence with Border Patrol U.S.-Mexico industrial development San Diego trolley to border Flooding Illegal immigration Sewage Tijuana River urban redevelopment Peso devaluation

Sources: *San Diego Union Tribune* and *Los Angeles Times*, 1950-1984.

Table 2.   Chronology of Newspaper Reporting on Planning Issues in the San Diego-Tijuana Region

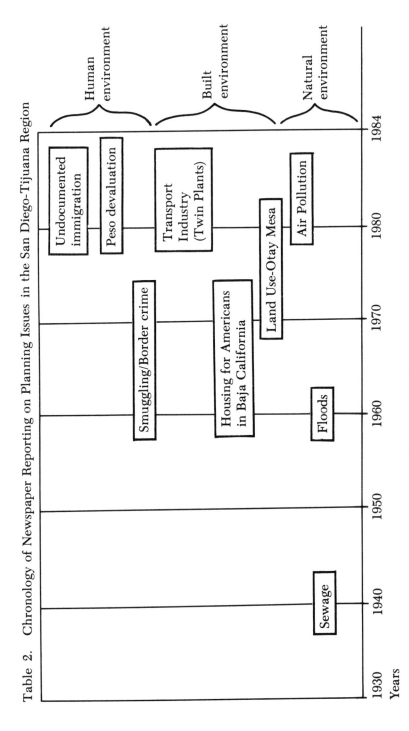

Sources: *San Diego Union* and *Los Angeles Times*, 1950-84.

1970s that externalities relating to the border built environment began to show up in local newspaper reports. Clearly this was, to some extent, a sign that the rapid urbanization occurring on both sides of the border in the 1960s began to generate built-environment externalities of a transboundary nature. By the 1970s, Tijuana began to undergo rapid transformation, some of these changes having had an impact on the San Diego region. Newspaper reports alluded to the redevelopment of the River Zone of Tijuana, the coastal development, the development of the Mesa de Otay, and the construction of a major industrial park housing factories that assemble U.S. products (Table 1).

The gradual overlapping of San Diego and Tijuana infrastructure signaled an important juncture in the history of the two cities. Prior to the 1970s the political boundary continued to separate San Diego and Tijuana, even though certain problems, like flooding or narcotics smuggling, concerned both sides. By the 1970s, however, the functional structure of the two urban centers had become integrated, so that future planning decisions automatically involved both sides of the border. In such areas as transport, housing, and industry the two cities began to fuse into a single functional metropolis.

The gradual evolution of a common bicultural built environment can be seen on two levels: direct material evidence in infrastructure and built facilities, and indirect examples of international influence in the cityscape. Direct material evidence is notable in various facilities. A second border crossing east of the San Ysidro border gate links land uses on the expansive Mesa de Otay/Otay Mesa development zone. Future land developments in this major growth corridor of both cities are likely to be more interconnected by the new border crossing. The River Zone of Tijuana, a new high grade commercial and office development zone of the city, displays considerable ties into the San Diego region through well-articulated freeway connections, bus lines, and extensive advertising in San Deigo county. Discussions of a possible international sewage treatment facility serving Tijuana and south San Diego represent another example of the possible blending of the built environment of the two cities, as does the gradual cooperation in linking U.S. corporations to the assembly plant program in Tijuana's new industrial park on the Mesa de Otay. Finally, the emergence of an "international housing market" gives evidence that housing markets no longer end at the border. Along the coast south of Tijuana are extensive housing developments occupied by American residents through long-term land leases with the Mexican government. These high-income housing submarkets within Mexico are matched by the emergence of pockets of high-income Mexican submarkets in south San Diego county, where wealthy Baja California residents have purchased housing in the U.S., although this pattern has been significantly moderated by the 1982 peso devaluation and the continuing struggles of the Mexican national economy.

Indirect evidence of a growing, symbiotic built environment exists as well. More specifically, we can now speak of the gradual "Southern Californiazation" of Tijuana, Mexico — the transformation of the city's built environment, over time, into one that is beginning to take on characteristics of a Southern California city. Tijuana's once concentrated structure has become more spatially extensive over the last decade, as the Mexican government has built new freeways and developed lands on the city's periphery. Where the urban elite once occupied properties near the center of the city, high and middle income suburbs are beginning to emerge on the outskirts of town. Housing styles in Tijuana have also begun to imitate U.S. architecture — the best examples are in the construction of condominiums in the River Zone and townhouses in the Chapultepec neighborhood. Some of the more recent high-rise buildings in the city use materials like glass and steel that seem to imitate U.S. high-rise structures. A good example is the recently completed twin tower Plaza Aguas Calientes office and commercial center complex. In short, Tijuana's gradual modernization and transformation illustrate a clear spillover of influences from San Diego's built environment upon the very physical structure and cityscape of this Mexican border metropolis. To a lesser extent are the visible signals of the spillover of Mexican culture onto the San Diego built environment. The appearance of Mexican murals on the pillars of the Coronado bridge and elsewhere is a case in point, as is the incidence of Mexican-Spanish architecture in a number of prominent public buildings.

Prospects for Management of the San Diego-Tijuana Transboundary Ecosystem

Given the striking empirical evidence of a transboundary ecosystem in the San Diego-Tijuana border metropolis, the question arises: to what extent can the ecosystem be adequately managed in the long run? The rapid demographic expansion of the cities of Tijuana, Mexico, and San Diego, California, is not likely to subside in the immediate future.

As the archival study of newspaper reports mentioned earlier demonstrated, problems associated with management of the built environment in the 1970s and 1980s have leaped onto the agenda of local concerns, especially those involving land use planning, transport management, sewage control, river channelization, housing, and industrial development. In addition, problems in the physical environment, especially those of air and water pollution, and social concerns such as crime and illegal immigration, continue to require binational attention. What then are likely scenarios for short- and long-term ecosystem management?

In considering management scenarios, it is essential to recognize that questions of planning for border cities in the future amount to the management of negative externalities outlined in the ecosystem model presented earlier. As Table 3 suggests, the most pressing concerns in the San Diego-Tijuana border zone involve the inability to make decisions cooperatively about the management of problems that have negative spillover effects on both sides of the border. These problems involve all three levels of the ecosystem modelled earlier — human, built, and natural environments. One very large problem at present is that most of the negative impacts of border problems are evaluated sectorally and in an ex post facto manner. In fact, little border land use or regional planning is taking place.[10] More important, evaluations and prognostications take place within each national jurisdiction, with very little transboundary cooperative analysis or research on San Diego-Tijuana planning problems transpiring presently.

In attempting to characterize the current status of transboundary ecosystem management in the San Diego-Tijuana region, a number of points are worth making:

1. Most decision making and negotiation is done through informal channels. With the exception of the International Boundary and Water Commission, which was established as a formal advisory and technical agency through international treaty, virtually no formal decision-making entities exist for dealing with border planning problems. The International Boundary and Water Commission has been successful in resolving boundary disputes, flood control problems, and questions of water quality regulation and water management.[11] As one document has stated, however, such ventures "owe their success to precise mandates and technical, essentially non-political modus operandi."[12] Nevertheless, many of the border problems involve political decisions. Because local officials fear insensitive intervention from the federal

government, they have tended to try to work things out through informal channels. A number of examples from the San Diego-Tijuana region corroborates this assertion. In the area of air pollution control, although a formal memorandum of agreement was signed between federal representatives in the U.S. and Mexico in 1978, local observers in the San Diego region claim that much of the activity and technical exchange has remained on an informal level.[13] Agreements regarding the location of the second border crossing, twin plant developments, transportation planning, police jurisdiction, and firefighting services in San Diego-Tijuana have all been reached through informal exchange. The planning for land development on the Otay Mesa/Mesa de Otay, probably the largest single urban growth corridor along the two-thousand-mile border, has been done on an entirely informal basis. No single transboundary document or formal instrument has emerged during more than a decade of studies, meetings, political maneuvering and, finally, drawing up of a master plan for land use on each side of the border. The planning processes have unfolded independently, and

Table 3.   Management Conflicts:
The San Diego-Tijuana Border Ecosystem

Problem	Nature of United States-Mexico Conflict
Flood control for Tijuana River	Mexico built concrete channel; U.S. opts for less compatible dissipator system
Control of sewage spills from Tijuana to San Diego	Unable to reach accord on emergency treatment procedures; planning for future sewage treatment remains unclear
Crimes against undocumented Mexican immigrants committed near boundary	Unable to resolve questions of police jurisdiction and cooperation near border
Management of air pollution	Unable to synchronize air pollution regulation and/or communicate between planners
Land use planning near the border	No mechanism for joint planning of such facilities as prisons, airports, industrial parks, hospitals, mass transit lines

in some cases conflicts have arisen. Informal dialogue has not always served to harmonize the interests of jurisdictions on each side of the boundary.[14]

2. All of the formal decision making, and much of the negotiation, that takes place is of a global nature; that is, communication is channeled through the federal governments of the United States and Mexico. The centralized nature of transborder coordination has been the subject of much discussion.[15] Observers believe that the principal forms of global decision making are treaties, presidential meetings, and inter-parliamentary conferences.[16] Again, the most significant treaty is the one that created the International Boundary and Water Commission in the late nineteenth century. Generally, however, local governments on both sides of the border have hesitated in placing the decision-making responsibility in the hands of the federal governments. As one inside observer has stated, "there is often the perception along that border that bureaucrats in the two capital cities are out of touch, disinterested, or not understanding of border matters."[17] A study of trans-boundary coordination of problems in the San Diego-Tijuana region clearly confirms the dominant role of the federal government. For instance, questions regarding the flooding of the Tijuana River and sewage contamination were handled by the International Boundary and Water Commission, the Environmental Protection Agency, and the State Department in the U.S., while the key Mexican agencies included the Secretary for Urban Development and Ecology and the Foreign Ministry, all Federal level agencies. Equally, the coordination of the second border crossing on Otay Mesa was handled by the Secretary for Human Settlements and Public Works in Mexico, and by the General Services Administration in the United States, both agencies operating in their national capitals.

3. Significant differences between the political organization of Mexico and the political culture in California severely hamper the ability of San Diego and Tijuana to cooperate in managing the transboundary ecosystem. Even when the two nations agree on the need to solve a given border problem, it is difficult to coordinate planning efforts both because the political systems are different, and because, as a result, both ignorance and misunderstanding linger about the workings of neighboring jurisdictions. Several specific differences in political organization seem especially problematic along the border and surface numerous times in the San Diego-Tijuana case: (a) centralized vs. decentralized federal systems; (b) stability of bureaucracies under

changing administrations; (c) the instrument of decision making; and (d) different goals in planning the border.

The Mexican political system is recognizably more centralized than that of the United States. Local governments have very limited input into decisions about the construction of highways, water lines, sewage facilities, schools, and other urban infrastructure in Mexico. The important decision makers are in Mexico City, first, and in the state capitals, second. Most public monies for development are handled at the federal level, so that local governments have very small budgets. While Mexicans are accustomed to this centralization, U.S. officials in San Diego government are forever bewildered by being unable to find their counterparts in Tijuana. For example, in city planning, housing, environmental, or transportation matters, the key official on the U.S. side is often in San Diego, while the key official in Mexico is in either Mexico City or Mexicali, the state capital of Baja California. The result is that communication between San Diego and Mexican officials is impeded simply by the disharmony between the political structures of the two nations.

A second major issue surfaces after bureaucratic liaisons are established between U.S. officials in San Diego and Mexican officials at the state and federal levels. U.S. bureaucrats have consistently voiced the frustration that when a new presidential administration (sexenio) begins every six years, the entire federal and state bureaucracy is virtually dismantled, rendering all previous contacts and transboundary lines of communication useless. This situation has particularly been frustrating for San Diego planners working in the areas of land use, air pollution regulation, and sewage management.

Another source of friction between the two political cultures lies in the differences between the instruments of decision making utilized on either side of the border. A number of empirical examples from the San Diego-Tijuana region illustrate this point. Mexican planning is wholly unique from planning in the United States. In San Diego, every community in the city of San Diego has its own master plan for land use, public services, environmental regulation, and growth management — these are called Community Plans. When U.S. land-use planners seek to understand the future growth of a specific subarea of Tijuana, they are frustrated to learn that no community plans exist. In fact, until 1985 no city plan for Tijuana had existed. Mexican planning tends to operate on an incremental basis. Different administrations will engage in specific projects according to the availability of funding. Some state governors have preferred to opt for building one memorable "monument" in Tijuana, to elevate their public image for the future. Comprehensive planning in Mexico is a very recent phenomenon and is still being slowly implemented in Tijuana. In the past decade "planning" consisted in government's taking control of certain lands through a federal land trust system (fideicomiso) similar to eminent domain in the U.S., then delegating the responsibility and funding for a specific type of development (industrial park, highway, river channel, public housing, etc.) to the appropriate federal agency. Plans would be drawn once the project was deemed appropriate by the federal government. This unique planning process and its instruments can be confounding to U.S. planners; the U.S. system may seem equally alien to Mexican officials. The point is that these differences contribute to an inability to effectively coordinate transboundary problems.

Not only may the instruments of decision making and planning differ; the goals of Mexican and U.S. border cities may also radically depart. Although it may seem rather obvious, we must remember that Mexico is at a very different stage in its economic evolution than is the United States; therefore, it is not surprising that the nations may have vastly different objectives along the border. In San Diego-Tijuana, a study of land use planning illustrated that even for land in the same growth zone on either side of the border, Mexican and U.S. objectives varied considerably. Mexicans were concerned about using the land to create jobs and housing and only *then* about bringing vital public services to residents, while the U.S. was more concerned with the environmental impacts of development and with resolving some controversies over the site planning of certain land uses that might threaten local property owners' ability to maximize their profits on the land.[18]

Toward a Transboundary Ecosystem Management Model

This essay has described the existence of a border ecosystem in the San Diego-Tijuana region, a functional habitat where the lives of several million residents from two cultures are linked across the international boundary. Empirical data gathered from newspaper archival sources indicates an increase in the number and types of transboundary linkages native to the ecosystem. More important, many of these linkages involve shared problems related to the gradual transformation of both San Diego and Tijuana into nationally ranked metropolitan centers. Rapid urbanization always demands adequate city planning techniques, as has been the case in Western Europe, the most densely urbanized region in the world. Rapid urbanization along an international boundary has sparked off an even greater challenge — bilateral management of a transborder ecosystem that is both heavily populated and growing.

A review of the record of cooperation among U.S. and Mexican officals in resolving planning problems in San Diego-Tijuana suggests that cooperative problem solving in the international ecosystem is in a precarious state. With the border region likely to continue its economic and demographic expansion, a mechanism for cooperative management of the border zone, especially in the large metropolitan regions, may represent one of the crucial items on the agenda of United States-Mexico relatons in the future. While not immediately evident, the visibility of border problems is presently overshadowed by two other regional problems receiving greater attention in the media — illegal Mexican immigration to the United States and political turmoil in Central America. In the not too distant future, it would not be surprising to see a swing toward national concern with the border region.

Clearly then, the order of the day is to begin to consider options for developing a management model for cooperation in controlling the problems shared by paired border cities. Recently, border scholars have made reference to the model of transboundary planning utilized in Western Europe and its possible application to the U.S.-Mexico border region.[19] The European model of transfrontier planning over the last decade has created mechanisms for dialogue, research, and creation of legal instruments to implement transborder coordination efforts. Most impressive are the number of transboundary planning entities that have been created along the borders of such international boundary zones as Germany-Netherlands, France-Switzerland-Germany, France-Switzerland, and Austria-Italy. In addition, the list of signed, interstate agreements between Western European nations is equally impressive, with the subjects of accord ranging from water pollution agreements, agreements over river navigation, and agreements on road links and air transport, to agreements on public health questions and frontier workers.[20]

The Western European transfrontier planning model owes its success to the existence of an interparliamentary structure, the Council of

Europe, capable of implementing the components of the planning model. It is obviously difficult to consider a direct transplant of the Western European model to the U.S.-Mexico border region, for many reasons. Aside from the unique history and cultures of the region, Mexico and the United States do not belong to any interstate parliamentary body similar to the Council of Europe. It should be added, parenthetically, that the Council of Europe probably owes its formation to the more powerful European Economic Community, a political body that emerged out of the recognition among Western European nations that their economies might be strengthened through cooperation. No such parliamentary structure is likely to form between the U.S. and Mexico in the near future. Given the history of each nation and the highly unequal state of their economies, little incentive exists on either side to form a parliamentary body oriented toward economic and trade matters.

What hope then is there for a cooperative model of border ecosystem planning? The key seems to lie in the resolution of two conceptual problems underlying the philosophy of a transboundary management entity: (a) the conflict between national sovereignty and a functional regionalist viewpoint; and (b) differences between national institutions involved in the border region.

The question of national sovereignty is fundamental to the United States-Mexico border zone. Although there is overwhelming evidence of an emerging symbiotic habitat, or ecosystem, this reality does not necessarily concur with what politicians on either side of the border might want to do from a symbolic standpoint. Mexicans are very proud of their heritage, and although the economic potential for trade along the border is acknowledged, they are very careful to protect the culture of the northern border. Recognition of the border ecosystem in a sense violates their strong sense of cultural pride and national sovereignty. The same might be said on the U.S. side of the boundary, except that Mexico, being the unequal partner, may feel even stronger about protecting its culture and soveriegnty. Yet these may be symbolic rather than tangible issues.

When we apply the question of sovereignty to the concept of the planning region, the conceptual problems are crystallized. Sovereign nations see regions, such as the border region, as "relay stations" subordinate to the primary national goal of economic growth. The sovereign nation viewpoint would not recognize that a functional border region may be so interconnected with the economy and geography of its national neighbor that formal regional planning ought to be implemented. Yet the notion of the sovereign nation is really a nineteenth-century concept; in an increasingly interdependent world, nations must reevaluate their thinking. In speaking of Western Europe, Quintin states that "the myth of national sovereignty only persists through ignorance of 20th century reality."[21]

The matter of divergent institutions along the border poses a second major obstacle to transboundary ecosystem management. If we again look to the Western European experience, we might note that many of the nation-states have common cultural heritages and philosophies about government, while this may not be true in the United States-Mexico case. Different institutions should not ultimately prevent cooperation, however; each nation can learn more about its neighbor's political structure once the premise of bilateral ecosystem management is agreed upon.

The real challenge, it would seem, will be for the United States and Mexico to come to terms with the symbolic tradeoff of perceived loss of sovereignty vs. the creation of mechanisms for formal transboundary cooperation. It would seem that what might be learned from the Western European model is that steps need to be taken to begin to set up mechanisms for border cooperation. Some of the pieces are already in place. What are needed are agreements to establish formal institutions for the exchange of knowledge and information and machinery that might act as a prelude to setting up model agreements on boundary problems that are particularly urgent. The sewage management problem in the San Diego-Tijuana region might be a starting point. The establishment of a transboundary working committee on sewage management would be a positive step towards the ultimate creation of a transboundary planning entity in San Diego-Tijuana.

Notes

1. U.S. House of Representatives and Senate, *Report of the Twenty Third Mexico-United States Interparliamentary Conference, Puebla, Mexico* (Washington, D.C.: U.S. Government Printing Office, 1984).
2. J.W. House, *Frontier on the Rio Grande* (Oxford: Clarendon Press, 1982), 251-54.
3. O. J. Martínez, *Border Boom Town, Ciudad Juárez Since 1948* (Austin: University of Texas Press, 1978).
4. P. de la Pradelle, *La Frontière, Etude de Droit International* (Paris: 1958), 56. Cited in J.M. Quintin, *European Cooperation in Frontier Regions* (Strasbourg: Council of Europe, 1973).
5. See C. I. Bevans, ed., *Treaties and Other International Agreements of the United States of America, 1776-1949*, Vol. 9 (Washington, D.C.: U.S. Government Printing Office, 1972).
6. J. Price, *Tijuana: Urbanization in a Border Culture* (Notre Dame: University of Notre Dame Press, 1973).
7. R. L. Bish, *The Public Economy of Metropolitan Areas* (Chicago: Markham Publishing Co., 1971).
8. S. Sayer, "The Economic Analysis of Frontier Regions," *West Europe Politics* 5:4 (October 1982): 64-80.
9. D. Harvey, "Labor, Capital, and Class Struggle Around the Built Environment in Advanced Capitalist Societies," in K. Cox, ed., *Urbanization and Conflict in Market Societies* (Chicago: Maaroufa Press, 1978), 9.
10. C. Graizbord, "Transfrontier Land Use Issues in the U.S.-Mexico Border" (Paper presented at the Conference on Urban Growth and Public Policy Options for the San Diego-Tijuana Border Region, University of California, San Diego, La Jolla, California, March 1985).
11. International Boundary and Water Commission, *Joint Projects of the United States and Mexico Through the International Boundary and Water Commission* (Washington, D.C.: Office of Mexican Affairs, U.S. Department of State, 1981).
12. P. Storing, "U.S.-Mexican Border Cooperation and Development: Is a New Border Commission Needed?" (U.S. Department of State, Washington, D.C., 1984), 3.
13. H. W. Brown and V. Bigler-Engler, "An Example of International Cooperation: The Tijuana-San Diego Air Quality Project" (Air Pollution Control District, San Diego, 1984).
14. L. A. Herzog, "The Cross-Cultural Dimensions of Urban Land Use Policy on the U.S.-Mexico Border: A San Diego-Tijuana Case Study," *The Social Science Journal* 22 (July 1985).
15. N. Hansen, "International Cooperation in Border Regions: An Overview and Research Agenda," *International Regional Science Review* 8:3 (1983): 255-70.
16. E. Mendoza Berrueto, "Visión de Un Marco Conceptual Para el Primer Encuentro Sobre Impactos Regionales de las Relaciones Económicas entre México y Estados Unidos," in A. Corona Rentería and L. James Gibson, eds., *Regional Impacts of United States-Mexico Economic Relations* (México, D.F.: El Colegio de México, 1984), 33-52.

17. Storing, "U.S.-Mexican Border Cooperation and Development," 2.

18. Herzog, "The Cross-Cultural Dimensions of Urban Land Use Policy."

19. Hansen, "International Cooperation in Border Regions." See also N. Hansen, "European Trans-boundary Cooperation and Its Relevance to the United States-Mexico Border," *Journal of the American Institute of Planners* 49:3 (Summer 1983): 336-43.

20. See Council of Europe, *The State of Transfrontier Co-operation Between Territorial Communities or Authorities* (Strasbourg: Council of Europe, 1982).

21. Quintin, *European Cooperation in Frontier Regions*, 9.

Part IV

The U.S.-Canadian Border

———

Recent Trends in U.S.-Canada Regional Diplomacy

Donald K. Alper

Overview of United States-Canada Relations

The United States-Canada relationship is probably unique in the world. No other two countries share such a long, undefended border. The degree of bilateral cooperation is unsurpassed, ranging from such vast joint projects as the building of the St. Lawrence Seaway to fighting side by side during World War II. Each country is by far the other's largest trading partner. The two governments' ability to manage common problems is generally viewed as the world's outstanding model of peaceful international cooperation. The degree and ease with which Canadians and Americans interact in cultural affairs, tourism, sports activities, and business affairs make it appear that the two societies are really one.

Yet close, friendly relations between Canada and the United States actually developed only in the last one hundred years. Relations between the two countries began amid suspicion; Americans broke away from the British colonial power while Canadians remained loyal. Indeed Canada served as a refuge for thousands of Americans who fled the Revolution, a fact which has led one Canadian historian to say that "Canadians are the world's oldest and continuing anti-Americans."[1] Although no military conflict between Canadians and Americans has arisen since the War of 1812 (a war that both sides claim they won), tensions between the two countries continued throughout much of the 1800s, fueled by the fear that America's "manifest destiny" might lead to northern as well as southern and western expansion. Canada's nineteenth century nervousness about American designs faded as both countries found themselves allies during the two world wars and as each economy became more and more connected with the other.

The concept of interdependence probably best captures the essence of the overall U.S.-Canada relationship during this century. In the economic realm, the sheer volume of bilateral trade is staggering. The total value of all trade between the two countries is some $700 billion. Roughly 70 percent of Canada's exports go to the American market, while Canada receives approximately one-fourth of all U.S. exports, making it by far the United States' largest customer. The two countries' military defense systems are linked through NORAD, NATO, and many other defense agreements. Corporate structures, professional associations, sports leagues, educational institutions, labor unions, and the mass media are all highly interlinked. Recognition of these and other areas of interdependence has led to a very high degree of cooperation and coordination between government and private officials.

The economic stakes in the relationship alone would suggest that both national governments would give the highest priority to the relationship. This is true for Canada. Ironically, the United States has in the past and continues to give U.S.-Canada relations very low priority; indeed American administrations pay far greater attention to smaller and more distant countries. This situation probably results from American officials holding the view that countries firmly within the U.S. sphere of influence that appear not to be susceptible to "communist" type insurrections do not require considerable attention. The official attention paid to Canada by the U.S. government is not only minimal but even embarrassing. It is revealing that Canadian affairs in Washington are conducted in a small Office of Canadian Affairs (within the Division of Europe and Canada) at the State Department. The U.S. Embassy in Ottawa is small and the ambassadorship carries little prestige. A former U.S. ambassador to Canada states that dealings with Canada are different from other countries in that they (Canadian relations) are handled in an *ad hoc*, dispersed manner by a variety of federal agencies with little or no regard to overall policy toward Canada.[2]

The major irritants in United States-Canada relations are transboundary environmental problems (especially acid rain); fisheries and maritime boundary problems on both the Pacific and Atlantic coasts; energy trade (especially hydro and natural gas); and, as of late, the

Canadian government's efforts to reduce the extent of outside owner-ship of major Canadian industries. What is unique about these irritants is that, with one exception, Americans perceive them as local or region-al problems and therefore unimportant in the general scheme of Ameri-can foreign policy. The exception is the foreign ownership issue, which has especially irritated the Reagan Administration since the Canadian government in 1980 proclaimed its National Energy Program designed to "patriate" a considerable portion of the oil and gas industry.

Although Americans find these problems of relatively little impor-tance, Canadians, most of whom live within two hundred miles of the U.S. border, view these issues as not only crucially important but in terms of national pride. This difference in perception has itself been a source of irritation. It has affected as well the development of cross-border problem-solving styles and mechanisms.

Subnational Transboundary Interactions

It could be said that the greatest number and most important in-teractions between the U.S. and Canada occur not between the two national governments in Ottawa and Washington but between subna-tional governmental units, private organizations, and individuals con-stantly interacting across the more than four thousand-mile border dividing the two countries. When one considers (a) that approximately three-fourths of the Canadian people live within 150 miles of the U.S. border, (b) that the traditional north-south regional ties (the Maritimes and Quebec and New England, Ontario and the midwestern states, the Prairies and the plains states, and British Columbia and the Pacific states) have created some regional interests, (c) that interest in Canada by Americans, except in the "border-belt" states, has never been very great, and (d) that common concerns, problems, and livelihoods have developed in the transborder regions, it is not surprising that subna-tional governments have increasingly taken the initiative to facilitate cross-border interaction and solve common problems.

According to a recent report by the Organization for Economic Co-operation and Development (OECD), all fifty states have some degree of interaction with the ten Canadian provinces.[3] This interaction takes a variety of forms. Many states have opened offices in Canadian prov-inces for the purpose of attracting foreign investment and trade oppor-tunities. The city of Seattle, Washington, on behalf of its electric pow-er utility, recently negotiated a draft treaty with the province of British

Columbia to settle the Skagit-High Ross Dam controversy. Several New England governors meet regularly with the premiers of eastern Canada and have established many agreements as well as an institutionalized forum for discussing problems common to all parties in the international region. Similar meetings are held by representatives of Alaska, British Columbia, and the Yukon. Although state and provincial international activity is becoming more widespread, by far the greatest amount of cross-border subnational contact is found between contiguous states and provinces.[4] Information flows (typically between administrators), understandings (a common result of "summit meetings" between governors and premiers), and formal agreements are rapidly increasing between subnational governments in the "border belt."

There is every reason to expect an increase in cross-border activity by subnational governments.[5] As noted earlier, the most troublesome issues in U.S.-Canada relations — environment, fisheries, energy, and agriculture — tend to be border-related and thus of primary importance to border states and contiguous provinces. Canadian provinces have shown an increasing inclination to interact independently of Ottawa and directly with their American neighbors on behalf of regional or provincial interests. Among the reasons for this are: the relatively powerful political position of the provinces in the Canadian federal system; geographical proximity imperatives — the cross-border movement of pollutants, people, energy flows, and economic activity on an increasingly larger scale; and continuing and perhaps increasing resentment, especially by Western provinces, toward a geographically distant and traditionally unresponsive central government. Similarly, American states, even though not as constitutionally powerful as Canadian provinces, are likely to increase their interaction with neighboring provinces. Calls for a "new federalism" and greater state self-reliance have prompted increased interest in cooperative regional initiatives in the areas of energy supply, freer trade across the border, marketing agreements specific to producers in a region, and even cross-border political alliances for the purpose of pressuring one or both federal governments.

Transboundary Institutions in State Government: The Case of Maine

For the most part, cross-border interactions between states and provinces are *ad hoc* and unstructured. The state of Maine provides an interesting case study because of its numerous interactions with three Canadian provinces and the fact that it has gone the furthest of any state in institutionalizing Canadian relations in the machinery of state government. Maine's interest in neighboring provinces is not surprising in light of its geography, history, and economic problems. Approximately two-thirds of Maine's land border is with Canada. More than 20 percent of Maine's citizens are of Canadian descent, and of this group, most have French Canadian roots, giving the state strong cultural ties to Quebec and parts of New Brunswick. Maine shares a common historical experience with the Canadian Maritime provinces, especially New Brunswick and Nova Scotia, and the economies all have a common resource base and suffer from similar problems — high unemployment, high energy costs, and insufficient financial resources.

A long history of personal contacts (governors and premiers) and administrative cooperation exists between Maine and the eastern Canadian provinces. In the early 1970s cross-border ties were strengthened when the governor of Maine and the premier of New Brunswick agreed to begin a dialogue that led to cooperative programs in the areas of the environment, trade, energy, tourism, and transportation.

In 1973 then-Governor Ken Curtis appointed a special assistant for Canadian Affairs and established the Office of Canadian Relations by executive order, the first and only such office in the United States. Its major function was to provide a full-time liaison between Maine and the eastern Canadian provinces. Such a liaison would not only facilitate communication and arrange visitations and other contacts, but it was expected the office would become the administrative focal point for dealing with routine cross-jurisdiction problems such as transportation coordination, disaster assistance, and pollution control. In addition, the office was given specific responsibility to strengthen all existing cooperative efforts with Canada; seek greater regional cooperation in the areas of economic development, energy supply, tourism, and environmental improvement; provide information about Canada-U.S. relations to the public; and encourage cultural exchanges and other modes of contact with eastern Canada.

Establishment of the Maine/Quebec-Maritime Advisory Commission to promote cross-border cultural and educational exchanges gave formal recognition to Maine's historic cultural ties to Francophone communities in the region.

In 1978 the Maine legislature decided to create its own Canadian relations structure which would be capable of responding to legislative concerns involving state-province issues and be independent from executive influence. A bill was passed creating a Maine-Canadian Legislative Advisory Office headed by a full-time director with fluency in the French language. A seven-member nongovernmental commission to assist the director in carrying out the mandate of the legislation was also established. It is significant that this legislation creating the new body referred to legislative action with legislative counterparts in Canada:

> There is established a Maine-Canadian Legislative Advisory
> Office which shall be concerned with strengthening all areas
> of regional cooperation between the Legislature of Maine
> and the legislative bodies of Maine's neighboring Canadian
> provinces. . . .[6]

In addition to promoting cultural activities and cross-border legislature exchanges, this innovative institution quickly became a vehicle for alerting legislators to border-related problems and facilitating solutions to them.

Executive And Legislative Roles

Maine is in the unique and sometimes awkward position of having institutional structures for handling relations with neighboring provinces in both the executive and legislative branches of government. This arrangement assures that border-related issues will inevitably find their way into the political process. However, "turf" problems have arisen as officials in both branches carefully guard their prerogatives in matters relating to the border and Canada.

On the executive side, the special assistant for Canadian Affairs (now called coordinator) is a member of the governor's personal staff, appointed by and responsible to him. The special assistant's chief function is to serve as a liaison for the governor in interactions with neighboring provinces and the Canadian federal government (through the Canadian Consulate in Boston). Other duties include troubleshooting cross-border problems which can be solved at the regional level, promoting tourism, and attracting Canadian business and investment into the state.

The Office of Canadian Relations has become an important support for Maine's executives in their dealings with provinces, but it has not replaced or even reduced the importance of personal contact and friendly relations between governors and premiers. Cross-border executive-to-executive relations have traditionally been extensive and exceptionally warm in the international region, with many recent governors and premiers having developed strong personal relationships.

The governor's personal role in Canadian relations is also enhanced because of the structure and operation of Canadian parliamentary government which is, to a much greater extent than in the United States, executive-centered. Provincial premiers, unlike their governor counterparts, do not have to contend with separate and formally "equal" legislatures in the conduct of intergovernmental relations. Premiers view executives, not legislatures, as their counterparts and engage in intergovernmental diplomacy accordingly. For example, in New Brunswick, relations with Maine at a level above department-to-department contacts are handled almost entirely by the Office of the Premier. An official in the Premier's Office told the author that in the Canadian system, serious issues of province-state relations are treated as matter of executive diplomacy in which premiers are the responsible actors.

The personal role of the governor in cross-border interactions is further enhanced by the New England Governors-Eastern Canadian Premiers Conferences, occasions for governors and provincial leaders to engage in summit-like diplomacy on major international issues in the region. Executives from six New England states and five eastern Canadian provinces meet annually to discuss problems and issues of mutual concern. The governors and premiers have been especially concerned with energy matters, regional trade, and tourism. While these meetings are sometimes criticized as expensive social gatherings, they have been very important in the development of a cross-border "we-feeling" among subnational elites.

On the legislative side, a Legislative Advisory Office was created in 1978. It is headed by a nonpartisan director who is appointed by the speaker of the House and the president of the Senate with the approval of a non-government commission consisting of academics, lawyers, and others interested in Canada and U.S.-Canada relations. The director of the Legislative Advisory Office works for the legislative members as a whole, much like a research officer or staff assistant.

The Legislative Advisory Office performs three major functions. First, the office performs a *constituency* function by assisting legislators with constituency-related problems involving Canada. As problems arise, the Legislative Advisory Office provides information and even acts as a kind of ombudsman on Canadian issues for constituents on behalf of the legislator(s). Approximately one-third of all Maine legislative districts border Canada; thus border-related problems are common. Examples of recent cross-border issues of particular concern to Maine legislators include spruce budworm infestation of forests in border regions, flooded potato farmers' fields along an international river caused by excessive amounts of water released by an upstream New Brunswick power dam, and minor, but politically sensitive problems such as working out reciprocity with Quebec for snowmobilers who venture across the mostly unmarked wilderness border. These kinds of problems are significant only to localized constituencies and fall naturally into the laps of state legislators.

Second, the office performs an *information* function by providing legislators and departments of state government data on a wide variety of topics related to Canada. The office assists the standing committees in the legislature by providing them with provincial documents, government publications, and other relevant information regarding Canada. The Legislative Advisory Office also assists departments of state government in their interaction with provincial governments by supplying background information, identifying Canadian counterpart agencies, providing translation services, and generally facilitating cross-border contacts. The office works closely with agencies involved in border-related problems such as the Public Utilities Commission, the Office of Energy Resources, the Department of Marine Resources, the Department of Environmental Protection and the Bureau of Forestry. Information is also directly disseminated through the issuing of occasional reports and brochures on specific Canadian issues.

Third, the Legislative Advisory Office promotes *interaction* between Maine and Canada through the vehicle of interparliamentary visits, conferences and cultural exchanges. For example, in 1982 a Maine-New Brunswick joint legislative conference was organized involving Maine legislators in the St. John Valley and their counterparts from border constituencies in New Brunswick to open lines of communication and discuss mutual problems in the region. In the cultural area, the office has promoted many cultural activities between the Franco-Americans in Maine, a sizable ethnic population, and the French Canadians and Acadians in Quebec and the Maritime provinces. To give one example, the office recently helped organize a "Canadian Festival" exchange involving high schools, performing artists, and linguists from Maine, Quebec, and New Brunswick.

Through its various functions — constituency, information, and interaction — the Maine-Canadian Legislative Advisory Office has not only increased legislators' awareness of Canada but also strengthened the competence of the state government to exercise influence over transborder issues. Maine's legislators are visibly active in Canadian affairs, a fact not lost in the halls of government in neighboring provinces. Yet a more structured relationship between legislators in Maine and counterparts in eastern Canadian provinces is not likely to develop given the basic differences in governmental systems. Legislative bodies in the provinces are subordinate to the cabinet and do not have the capacity to act independently as do their counterparts in the states.[7] This suggests that province-state interactions will be centered on executives and that predominant authority to speak on behalf of the state and provinces will continue to reside in the governor's and premier's offices.

Assessment of the Maine Case

What conclusions can be drawn from Maine's institutional framework for interacting with neighboring provinces? First, Maine's state officials in both the governor's office and the legislature have relatively open channels of communication with counterparts in the neighboring provinces. International borders have a way of blocking access of one subnational government to another. State officials need to have access directly into provincial decision making, particularly important because of the powers provinces wield in the resources area. Former governor Ken Curtis credits close, state-provincial communications for preventing the building of an unwanted oil refinery in an environmentally sensitive area in southeastern New Brunswick near the Maine border and for making possible the negotiation of a special oil export agreement with New Brunswick to avoid the shutdown of a Maine paper mill and the laying off of hundreds of workers.[8]

Second, the institutionalization of Canadian relations in both the executive and legislative branches of state governments strengthens the state's capacity to respond to border-related problems as well as to engage in cooperative regional endeavors with counterparts in the neighboring provinces (energy communities, cooperative tourism promotion, regional economic development, etc). Border-related problems weigh disproportionately on certain constituencies, and the existence of a Maine-Canadian Legislative Advisory Office in the legislature is an aid to their resolution. Similarly, Maine governors are better able to speak and act on behalf of the state's interests in summit conferences and regional international meetings.

Third, Maine's institutional innovation in Canadian relations has better equipped the state to play a stronger role vis-à-vis both federal governments in Canadian-related problems and concerns. The structure of American federalism distributes power unequally, with small states like Maine having relatively little power in the federal system, which in turn places them in a weak position to affect federal actions as they relate to U.S.-Canada relations.

Much of Maine's institutional innovation in Canadian relations should be understood as a means to better equip the state to play a stronger role in specific Canadian-related problems and concerns. The fact that subnational governments have unequal power in the federal system fosters regional strategies as a way of increasing political clout against the federal governments in both countries. The Maritime provinces find themselves in a similarly weak position vis-à-vis the Canadian federal government and thus similar regional strategies are found in Canada. Cross-border cooperation is a natural way to increase the overall power of Maine and its neighboring provinces to affect U.S.-Canada relations. In testimony before a Canadian Senate committee studying state-province interaction, former Governor Curtis of Maine summed this up:

> I think one of the advantages of regionalism . . . is that there is an opportunity to get together as a region and discuss these barriers to what might be obviously good for the two areas, and then direct some kind of formal voice to the respective federal governments expressing this problem.[9]

A similar view was recently expressed by Washington State Governor John Spellman, during a visit to the British Columbia Legislative Assembly, when he called on the province to join Washington and other northwest states to pressure collectively the U.S. and Canadian federal governments to reach an agreement on a salmon treaty for Pacific coast waters.[10]

Federal Involvement in Subnational Interactions

The Maine case is an example of an institutional framework which strengthens the state's capacity for handling cross-border interactions with neighboring provinces. Most transboundary problems between Canada and the United States are viewed by the respective national governments as not important enough to require attention, or they are seen as a part of a much broader process of bilateral negotiation in which the local problem often gets used as a bargaining chip for achieving other objectives. Subnational governments thus feel a need to deal directly with their cross-border counterparts, not only because of the relative lack of concern of the federal governments but, ironically, because the respective federal governments too often *do* get involved.

Issues involving fisheries, local pollution, energy, and agriculture, because they tend to be more naturally local or regional than national, have growing sentiment that they can be handled more effectively by local officials and interested parties. Too often it is the case that once the national governments elevate a primarily local issue to a place on the U.S.-Canadian diplomatic agenda, it becomes a matter of national importance and thus a new context develops where it is mixed into the national interests of the two countries. With new actors involved, the issue typically gets politicized — increased attention to the issue by government officials and interest groups — and resolution becomes more difficult, even more a problem when high government officials "link" unrelated issues.[11]

The U.S. federal government has recognized in fact if not in theory that it is neither possible nor even desirable to deal directly with the vast array of issues arising out of U.S.-Canada interactions. What seems to be evolving is a joint or collateral involvement of local, state, and federal government actors depending on the issues and the degree to which issues get politicized. Federal governments in both the U.S. and Canada are increasingly reluctant to get involved in local cross-border matters unless a fairly substantial national interest (or point of national pride) is involved.

Neither government has made a particular issue of subnational cross-border diplomacy and both seem content merely to try to keep track of what is going on. In both the U.S. State Department and the Department of External Affairs, officials monitor the interactions between states and provinces. Career foreign service officers from time to time work with the various Governors' Conferences and the Council of State Governments. When regional summit conferences are held between governors and premiers, observers from Ottawa and Washington regularly attend. Administrative offices such as the U.S. Trade Representative, the Commerce Department, and the Environmental protection Agency remain in close contact with state officials.

The Role of the International Joint Commission

The one federal institution that has regularly intervened in local cross-border issues is the International Joint Commission, created under the Boundary Waters Treaty of 1909. Comprised of an equal number of commissioners from each government, its purpose is to investigate, examine, and recommend with regard to changes in rivers, lakes, and waterways which cross the border. The IJC has no enforcement powers and thus cannot commit a country to a course of action. Its historical role has been that of "independent" fact finder and arbiter of crossborder differences before they escalate into major conflicts. The commission has had notable success in allocating transboundary waters between upstream and downstream users and providing the impetus for cleaning up the Great Lakes. It was largely responsible for creating a management scheme for the Great Lakes-St. Lawrence waterway system and continues to administer the various Great Lakes Water Quality Agreements.

It is commonly believed that the IJC is the ideal mechanism for handling contentious cross-border local and regional issues. This view is based on the IJC's reputation as a fair and generally nonpolitical international organization. However, the actual work of the IJC reveals its limited capabilities. First, the IJC can only act if both governments have referred a case to it. Second, although there is literally no limit to what may be requested from the IJC as regards the universe of U.S.-Canada relations, the organization's traditional role and major experience is in the area of transboundary water changes and pollution. References in other areas, such as bilateral trade or cultural disputes, would probably require a fundamental change in its organization and operation. Third, the IJC has performed most of its successful work in politically noncontroversial areas where both sides have shown a strong interest in studying and resolving the problems. In recent years most of the commission's opinions in controversial cases have resulted in nonbinding recommendations which one or both governments have simply ignored.

The IJC cannot resolve transboundary problems unless the involved governments find it in their interests to accept the commission's findings as binding. The basic problem is that the IJC has been mostly a fact-finding and mediation body — although a highly respected one — which must contend with the numerous governmental jurisdictions and public and private interest groups which have an interest in transboundary environmental problems. As long as everyone is interested in cleaning up a transboundary river or lake, the IJC can be an effective facilitator because of its scientific work and ability to focus attention to the issue and bring the disputants together. Most of the IJC's successes have occurred in these kinds of "non-zero sum" situations. The Great Lakes project is an example of this; pollution costs and cleanup benefits were shared on both sides of the border, and it is a water system that would have suffered serious pollution regardless of whether it was wholly contained in one country or part of two. Curtis and Carroll argue that the Great Lakes have not suffered by being shared by the U.S. and Canada; to the contrary, they have received increased attention because of their international status and are probably better off today because of it.[12]

Unfortunately, most cross-border environmental and energy issues tend to arise because costs are disproportionately imposed on one side of the border while benefits are produced on the other. There are a number of recent cases like this: The High Ross Dam-Skagit controversy (recently settled) involved the raising of a dam in Washington State which would flood thousands of acres of wilderness land across the border in British Columbia; the Cabin Creek Mine controversy between B.C. and Montana stems from the proposed building of a huge coal mine in southeastern British Columbia which would have a serious water and air pollution impact on prime recreational and tourist areas in Montana's Glacier National Park; the Poplar River Project in Saskatchewan entails a large power plant, strip mine, and dam on the Poplar River which crosses the border into Montana thereby threatening restriction of water supply for downstream agriculture and pollution damage to the fish and the waterfowl; the Garrison Diversion is a major manipulation of the waters of the Missouri basin in North Dakota for irrigation purposes which would divert waters containing noncompatible minerals and fish life north into the Souris-Red River-Hudson Bay drainage in Manitoba; the Eastland Oil refinery controversy on the Atlantic coast has involved an attempt to locate a large oil refinery at Eastport, Maine, just across the New Brunswick border, and thus would have placed the New Brunswick coastline and economically important fishery in jeopardy from passing oil tankers.

Collaborative Involvement Among Governments: The High Ross-Skagit Case

The recent settlement of the High Ross Dam-Skagit case between Seattle, Washington, and the government of British Columbia suggests a less direct, but potentially significant role for the IJC as an ancillary, yet subtle political force in a dispute which involved subnational governments. In 1981, after years of contentious argument between Seattle (on behalf of the city's electric utility, Seattle City Light) and the province of British Columbia over whether to raise the Ross Dam and flood B.C. wilderness area upstream, a negotiating team was created under the auspices of the IJC to try to resolve the problem once and for all. The team was composed of officials from Seattle, B.C., and one member each from the U.S. and Canadian commissions of the IJC. Although the IJC had been long involved in this issue, past negotiations had resulted in stalemate.

In 1983 a breakthrough leading to a settlement occurred. The city of Seattle agreed not to build the dam in exchange for the province's agreement to sell the city an equivalent amount of power at an equivalent price until the year 2066. The city would pay British Columbia the amount of money it would have spent raising the dam.

The IJC performed none of its traditional roles — adjudicative, investigative, administrative, or arbitrative — in this case; instead it exercised political influence. In effect, the Commission told both sides if they did not work out some kind of settlement, the IJC acting on behalf of both federal governments would recommend its solution. This kind of political prodding is not characteristic of IJC practices in the past, but the IJC itself has changed in the last few years. What is different is the relative position of the IJC in the Washington bureaucracy. It is significant that President Reagan appointed three new commissioners in 1981, all conservative Republicans with close personal ties to the president. This ideological base will mean an increased politicization of the body, but it is likely that the commission henceforth will have a greater capacity for action because of its ties to the White House. Schwartz and Jockel conclude that the commission, because it is more politicized, has come to be an important actor in Washington, "a lobby from within," and therefore a potentially more powerful force in promoting resolution of U.S.-Canada transboundary issues.[13] The High Ross Dam case suggests the IJC can be useful in helping to resolve regional cross-border issues if its role is that of providing leverage on behalf of both national governments while local parties to disputes take the lead in negotiating settlements.

Lessons/Conclusions

Subnational cross-border interactions are not likely to decrease in volume in the future; if anything, they will probably increase. Given the unusually high degree of interdependence between Canada and the United States, and especially between states and provinces in the border region, subnational governments are driven to cooperate and interact to meet economic, trade, environmental, communication, transportation, and health imperatives.

Can we expect more border states to follow the Maine example of establishing institutions in state government for handling problems and relations with Canada? Although this appears to be desirable in states like Washington and Montana, certain political factors weigh against greater formalization. The different governmental systems and political cultures make official interactions and institutional (especially legislative) ties difficult. A conventional view among state officials is that cross-border interactions are essentially bureaucratic and not important enough to warrant a separate agency level office or even a high-ranking advisor. States have also found provinces to be less interested in formalized cross-border ties which might limit freedom of action and create the symbolic effect of being formally tied to the United States. Most provinces are considerably larger than contiguous states and, with exclusive ownership of their resources and considerable power in international trade, find that they have cross-border interests which go beyond their immediate U.S. neighbors. Quebec's U.S. "hinterland" includes all the New England states plus New York and other mid-Atlantic states. Alberta views relations with Colorado and other Rocky Mountain states as primary in addition to Montana.

British Columbia's economic/trade ties with California are far more important than all of the Pacific Northwest states combined.

It appears that some kind of institutionalized Canadian relations capacity in state governments is desirable to alert and educate government officials about border-related issues and provide channels for official contact, deliberation, and intelligence to handle state-province cross-border initiatives and resolve problems. When transboundary disputes occur the "normal" inclination is to seek recourse from the federal governments which frequently results in exacerbation of the problem or prolonged negotiations in which the local concern becomes part of the U.S.-Canada issue agenda. Maine has gone the farthest in giving central organizational expression to its interactions with Canada, but other border states are increasingly finding it necessary to strengthen their border relations capacity. In 1981 Montana created an *ad hoc* committee on Canadian relations composed of agency heads to provide ongoing monitoring of Canadian events and issues. In 1977 Washington State created a joint Senate-House committee called the Joint Legislative Committee on Washington-British Columbia Cooperation which was to be a vehicle for legislative interaction with the British Columbia parliament. In Michigan and Vermont the governors have taken steps to equip themselves better to interact with neighboring provinces.

Perhaps the greatest success in managing cross-border problems is found in regional institutions like the joint governors-premiers meetings. Regional approaches are especially appropriate considering the fact that many of today's most important, and perhaps most troublesome, issue areas in U.S.-Canada relations involve and affect multiple political jurisdictions on both sides of the border (e.g., acid rain, fisheries, energy, tourism, and regional economic development).

The New England Governors-Eastern Canadian Premiers Conference, most successful of the cross-border regional meetings, laid the foundation for a transnational energy community and, in an innovative project, established a central mechanism for region-wide promotion and coordination of tourism in the New England/Quebec/Maritimes area. The cross-border regional approach has been followed in the Great Lakes area where Midwestern governors and neighboring premiers have cooperated in the cleanup of the Great Lakes. On the west coast, annual meetings are held between officials from Alaska, British Columbia, and the Yukon Territory. At the 1983 Western

Premiers' Conference considerable interest was evidenced in a future meeting of western premiers in conjunction with their governor counterparts from neighboring states. Among the topics suggested for such a meeting was a new western hydropower grid and strategies for coordinating opposition to federal resource policies thought to be too intrusive and restrictive. There is also growing interest in cross-border regional economic development to increase marketing opportunities and productivity by integrating economic activities in adjacent areas which, because of the international border, are now competitive. The Canada-U.S. Auto Pact and the proposed twelve-mile border free-trade zone between the U.S. and Mexico are examples of this approach.

One of the most important lessons that emerges from Canada-U.S. relations is the importance of regional/local authorities negotiating local issues whenever possible. Not only are local issues removed from their context when elevated to a higher place in U.S.-Canada relations, but they will likely fall into the hands of negotiators who neither fully understand nor even necessarily share the local actors' concerns. What is fair for both national governments may not be fair or even acceptable to Seattle or Alberta or Maine border towns on the New Brunswick border. Federal management and control also is perceived as interference with the right of Canadian provinces to pursue foreign policy activities on their own and, to a lesser extent, states' rights in the United States. For constitutional as well as socioeconomic reasons, province building in Canada has become the most significant feature of Canadian federalism and most provincial premiers view their role as akin to running quasi-sovereign nations. Provincial prerogatives and "rights" are jealously guarded, as was seen during the Skagit-High Ross Dam issue when B.C. steadfastly maintained a strong provincial position and insisted on negotiating its own settlement.

Although certain areas do exist in which only federal authorities have the appropriate economic and/or constitutional powers and authority, both federal governments have recognized the folly in trying to manage centrally such a complicated relationship. If there is any trend in transborder regional diplomacy, it is that we can expect a greater degree of collaborative activity involving all levels of government in the future conduct and management of U.S.-Canada relations. The recent Skagit-High Ross Dam case provides a case in point; local officials representing the city of Seattle and the province of British Columbia reached a settlement with the prodding of federal officials who exerted leverage through the IJC. What is all the more remarkable is not that the outcome was acceptable to the two national governments (the treaty is before the Senate as of this writing), but that both Canada and the U.S. encouraged through the IJC the two most locally involved parties in the dispute — Seattle and B.C. — to negotiate an international agreement in the place of the two governments' formally constituted diplomatic authorities in Washington and Ottawa.

Notes

1. Ronald G. Landes, *The Canadian Polity* (Scarborough, Ont.: Prentice Hall of Canada, 1983), 340.

2. Kenneth M. Curtis and John E. Carroll, *Canadian-American Relations: The Promise and the Challenge* (Lexington, Mass.: D.C. Heath and Co., 1983), 10.

3. Ibid., 66.

4. Roger F. Swanson, *Intergovernmental Perspectives on the Canada-U.S. Relationship* (New York: New York University Press, 1978), 250-51. According to Swanson's study, the fourteen border states account for more than 60 percent of all state interactions with Canada. States in the Northeast region were involved in more than one-third of all interactions and the most active state-provincial pair was Maine-New Brunswick followed by Michigan-Ontario and Washington-B.C. respectively.

5. Ivo D. Duchacek, "Transborder Regionalism and Micro-Diplomacy: A Comparative Study" (Paper presented to the Seminar on Canadian-United States Relations, Harvard University, 6 December 1983); Donald K. Alper, "Transnational Regionalism and Canada-United States Relations" (Paper presented to the Western Social Sciences Association Convention, Albuquerque, N. Mex., 29 April 1983).

6. State of Maine, *Main Revised Statutes Annotated*, 1964, vol. 2, title 3, sec. 221-228 (Saint. Paul, Minn.: West Publishing Co., 1979): 198.

7. A detailed discussion of this problem is found in Gerard F. Rutan, "Legislative Interactions of a Canadian Province and an American State," *The American Review of Canadian Studies* 9:2 (Autumn 1982): 67-79.

8. Curtis and Carroll, *Canadian-American Relations*, 70.

9. Senate of Canada, *Proceedings of the Standing Committee on Foreign Affairs* no. 7 (20 February 1975), 7, 10.

10. *Bellingham Herald*, 15 February 1984, 2A.

11. Robert O. Koehane, and Joseph S. Nye, "Introduction: The Complex Politics of Canadian-American Interdependence," in Annette Baker Fox, et al., *Canada and the United States: Transnational and Transgovernmental Relations* (New York: Columbia University Press, 1976), 13.

12. Curtis and Carroll, *Canadian-American Relations*, 29.

13. Alan M. Schwartz and Joseph T. Jockel, "The International Joint Commission and Great Lakes Water Quality: An Emerging Lobby from Within?" (Paper presented to the Association for Canadian Studies in the United States, Annual Meeting, 30 September 1983).

The Subtle Invasion:
Canadian Cultural Transfers
to the United States

Victor Konrad

In Bellingham, Washington, with Canada's third largest city Vancouver a mere hour away, a turn of the television dial reveals Harvey Kirk and the decidedly Canadian viewpoint of the CTV news — understated television drama on the CBC and federally subsidized French language programming. For most Washington residents, Canada's French culture leaves only a momentary impression, not fully understood, but along the northeastern boundary it is a cultural lifeline for Franco-Americans residing in upstate New York and northern New England's tier of Vermont, New Hampshire, and Maine. Here language serves to transfer culture between related minorities. In Washington and states west where English predominates along the border, a subdued Canadian message permeates the seemingly solid front of American programming beamed across the border. Canadian views, personalities, and ways are assimilated, often unconsciously, by neighbors in the United States. Canadian cultural transfer is apparent and undaunted in the face of United States cultural dominance in English-speaking North America.

A subtle invasion of Canadian culture is taking advantage of the inclusive nature of American culture. Over several centuries Canadians have learned that the United States folds other cultures into a constantly changing American mix. In contrast to the Canadian mosaic of founding British and French,[1] European and Asian immigrant enclaves, and world refugees who now call Canada's cities home, the

United States symbolizes cultural anonymity in its blend, a blend which exudes popular and vernacular culture at a national scale.[2] The American popular culture barrage to Canada is continuous, for it finds a market and even some empathy there. Now Canadians block some American incursions and filter the rest for Canadian consumption. Meanwhile, elements of Canadian culture — art, film, music, sport, writing, architecture — cross the border to be blended in Hollywood and New York, sustained in St. Petersburg or Tucson, or adjusted in small communities from Calais, Maine, to Port Angeles, Washington. The paradox of the Canada-United States boundary is that the border serves not the dominant state so much but more so the lesser state. Always an active conduit for flows of capital, goods, and lifestyles from the United States to Canada, the border increasingly assures Canadian identity while allowing Canada to export to the United States and Canadians to sustain elements of exclusive Canadian culture in the United States.

Models of Cultural Transfer

Canada shares its land borders only with the United States; Arctic water and ice boundaries are expansive and in effect barriers to cultural transfer from circumpolar nations. This condition simplifies applying and testing models of cultural transfer, since transoceanic influences may be isolated and separated from consideration of borderland interaction.

When Britian left the United States in the late eighteenth century and concentrated its forces in the country newly won from France, a situation of opposed armed camps prevailed in North America. Distances between the independent states on the eastern seaboard were considerable, however, and isolation of developing national entities prevailed on both sides of a largely unoccupied border zone. During the next century, settlement areas grew to meet first at border point frontiers on the Gulf of Maine, the Lake Champlain valley, the Niagara Peninsula, Detroit, and the Sault. Attempts at conquest in the early nineteenth century resulted only in confirmation of the boundary, and the young United States proceeded to establish a different border relationship with Britain and subsequently with the new Dominion of Canada. Conquest and cultural subjugation[3] never occurred, and a border landscape of adjacent but distinct cultures was not realized (Table 1). Neither was the United States able to impose its newfound identity on its Canadian neighbor. Instead, transformation[4] developed early in the relationship through trade that grew constantly until each of the neighbors became each other's major trading partner. For the more populated and highly industrialized United States, a clear advantage emerged as the U.S. economy took over from Britain to dominate in Canada. Canada moved from a situation of trans-Atlantic mercantile control to one of adjacent economic control through American branch plants. Automobile manufacturers, oil companies, chemical industries, appliance manufacturers, and even food producers transferred plants, labels, and management across the border and tied Canadian industry and business into the American industrial complex.[5] American culture merely rode the wave into Canada. Where the United States presence was strong in cities, mining centers, and manufacturing locations, American culture saw its greatest expression. Since most Canadians lived (and continue to live) within two hundred miles of the United States border, the impact of cultural transfer was concentrated and magnified; Canada's heartland[6] and borderland were the same.

Table 1. Models of Cultural Transfer Across
          International Boundaries

Theory	Process	Border Landscape
Conquest	Subjugation	Contrast/Imposition
Transformation	Trade	Transplanted Developments
Accommodation	Adjustment	Coalescent Features

With greater realization of identity after the 1967 centennial, transplanted U.S. developments were tolerated less and Canadians embarked on a decade of national and regional expression. Whereas increased cultural transformation marked previous decades, accommodation of Canadian and American cultures emerged in the 1970s to herald a new form of transfer. Canadians adjusted to American cultural dominance by limiting U.S. news magazines and restricting employment of American professors in Canadian universities while increasing support for domestic cultural agencies and extending efforts to convey the Canadian perspective in the United States.

Such adjustment already is showing signs of coalescent features in cultural landscapes of the border country. If accommodation continues, the borderlands may reflect increasing evidence of a transition zone between Canada and the United States.[7]

The current model of cultural transfer does not apply equally along the four-thousand-mile boundary spanned by at least six distinct regions of cross-border interaction. However, in all these regions some form of accommodation is in evidence and confirmed in the border landscape. Perhaps because it is a multilevel form of cultural transfer, one which allows adjustment at local, regional, and national levels, accommodation now supercedes cultural transformation established through trade and regulated at a national level.

In the complex but relatively similar cultural experiences of the United States and Canada, neither conquest nor transformation proved viable means of cultural transfer between the two countries. Instead, accommodation, which recognizes separate identities from local to national levels and allows cross-border transfers, had evolved as a model for explaining contemporary Canadian-American cultural relations.

American Culture in Canada

American culture predominates English-speaking North America. It predominates to the extent that most people in the United States do not recognize that another distinct cultural force exists and indeed thrives north of the border. American culture in all its facets has impacted Canada tremendously. The aspects studied by cultural geographers — expressions of ethnicity, language, art, agriculture, architecture, folkways, and popular culture — all show the imprint of United States influence and presence. The impact, although somewhat diminished in recent years, remains in the media flow across the border, the transfer of American culture to Canada by people from the United States, and the baggage of American culture brought home by returning Canadians.

The imprint is visible on the land. Vancouver's skyline repeats San Francisco, Portland, and Seattle, and on the surface urban landscapes also appear similar. California's influence is seen on building murals and in British Columbia's adaptations of the California bungalow found from upper class residential areas to working class neighborhoods.[8] Calgary's oil company towers identify a northern Houston influence, and Cardston's Mormon temple is virtually the twin of a structure which dominates Mesa, Arizona. In Winnipeg billboard design imported from the United States drapes buildings along with distinctly Canadian advertisements. Ontario's auto styling remains consistent with United States standards and continues a tradition of cross-border influence which shows roots in the early nineteenth century, when Federal design found its way across the border to the heartland's buildings. Tides of housing taste expressed in the United States are replicated in Ontario, and even the American flirtation with octagonal houses and barns is found there. In Toronto gentrification of declining central city neighborhoods mirrors the experiences of Philadelphia, Baltimore, and other old eastern cities where professionals have returned to the urban core to live. As in U.S. cities, the symbolic relationship to the past is embellished in refurbished housing, reconstructions, and the new made to look old.[9] The celebrated past may differ from city to city, but the process remains the same.[10] Canada's Maritimes show both the new American influence of oil culture and the old Cape Cod houses brought from Massachusetts and accepted in the Canadian portion of the Atlantic region.[11] A virtual landscape of American culture prevails across Canada and remains a consistent and repeated element in the Canadian cultural mosaic.

Canadian Identity, the Land, and Cultural Transfer

Canada's identity originally was based on the land, a northern land of illusion, a land which few Canadians ever saw. It was a northern identity rooted in visions of Mounties, Eskimos, and translated to the south as the television series "Sergeant Preston of the Yukon" or the films immortalizing the Inuit hunter "Nanook of the North." Although these visions are history to Canadians, the American public retains an image of Canada as painted by the "Group of Seven," a Canada of northerliness, of Black Spruce tangle, musk-oxen, and Canada geese.[12]

The image is sustained when Americans who venture north see the strong native presence and notice frontier landscapes. Oppressive forests, mountains so familiar they find their way onto whiskey bottles, glacial survivals, and severe winter scenes remain symbolic.

For Canadians, centennial fervor helped to mold a new image, and with Trudeaumania, a new and distinctive Canadian flag, and pervasive federalism, Canada began to etch a different identity at home and abroad. It is an identity which allows strong nationalism and strong regionalism both to prevail simultaneously; English and French are celebrated and acknowledged as official languages and founding cultures; other cultures also remain exclusive in confederation. In spite of a Quebec move toward sovereignty, western concerns about eastern political and economic domination, Newfoundland's claim to a more equitable share in confederation, and northern drives toward autonomy, Canada remains intact and vigorous as a nation.

Seemingly off balance and in turmoil, Canada has managed to muster a subtle invasion of the United States. The undefended border which allowed transfer of U.S. culture to Canada also allowed Canada to export culture to the United States. The shared borderland, the land *Between Friends* (National Film Board of Canada), has become more than a boundary along which coincident natural features like Niagara Falls and joint historical legacies like Champlain's initial settlement on Dochet Island in the St. Croix estuary mark a common experience. As illustrated in *Between Friends*, Canada's bicentennial gift to the United States, the borderland has emerged as a set of regionally distinct transitions between Canada and the United States. Beyond the borderland stand enclaves and expressions of Canadian culture to proclaim the vitality of cultural transfer to the United States.

Illustrations of a Subtle Invasion

Before providing illustrations of transfers, it is necessary to emphasize the importance of the Canadian retaining wall in the process of cultural adjustment along the border. Government demands for high Canadian content levels in news magazines, radio and television broadcasting, and subsidized publishing helped to establish a barrier behind which Canadian film, art, crafts, sport, and letters also are nurtured. Meanwhile, the barrier has restricted U.S. cultural incursions by filtering or reducing wholesale American transfers which threatened the viability of fledgling Canadian cultural institutions and private sector initiatives. The *Time* magazine case provides a good example. *Time* refused to enhance its Canadian content in the edition published for sale in Canada and challenged a Canadian government order to comply. The order stood and *Time* dropped its Canadian edition entirely. Canadians reacted by reducing subscriptions drastically and opening the way for *Maclean's*, a former monthly, to become Canada's weekly magazine.

The cultural retaining wall and growth of Canadian culture at home soon spawned an increase of exports to the United States, exports that followed previously established paths to familiar destinations. This surge of Canadian presence was evident both at the border and deep in the United States.

Annual migrations of Canadian "Snowbirds" to Sun Belt states increased during the last decade and furthermore established visible communities from California to Florida. Winter visitors and particularly retirees form distinct enclaves identified by Canadian flags and commercial signs designed to lure the dominant Canadian clientele. Québecois communities near St. Petersburg and Clearwater, Florida, support French language newspapers and attract small business entrepreneurs from Quebec to "Le Sud" in order to cater to their needs.[13] The rise in mobile home use has increased rather than decreased community cohesion, for mobile homes easily regroup to former stronger enclaves in newly discovered Sun Belt locations in Arizona, Georgia, and South Carolina.

A well-established enclave in the United States, the Hollywood community of Canadian film-makers and actors, now serves as one terminus rather than the destination for Canadian talent. Toronto has emerged as "Hollywood North" and attracts American as well as Canadian film and music enterprise. From the Toronto publishing houses of McClelland and Stewart, Anansi, and many smaller presses, the works of celebrated Canadian writers like Margaret Atwood, Farley Mowat and Margaret Laurence are finding considerable distribution in the United States. Added to these exports are a burgeoning array of children's books with content which conveys Canadian land and life.[14] On the American side of the border, Canadian Studies institutes at universities from Maine to Washington act as cultural diffusion points as well as centers for the academic study of Canada. The vigor of Canadian Studies in the United States is evidenced by a national, interdisciplinary organization, the Association for Canadian Studies in the United States, which now boasts a membership of more than one thousand, academics, government and private sector professionals, and teachers. Their journal, *The American Review of Canadian Studies*, more than a decade in print, was joined in 1984 by the *Journal of Canadian Culture* as the second scholarly publication on Canada originating in the United States. Similarly, Canadian native arts and folk art traverse the border duty free and are drawn into the massive U.S. market for these treasures. Northwest coast motifs from Haida, Tlingit, Nootka, and Salish culture are now incorporated into the design of American Plains and Southwest native artists, these incorporated forms sold beside both traditional Northwest and Southwest native art. Add to these cultural transfers popular elements such as the "Moosehead Madness" of beer and associated regalia originating from a small brewery in New Brunswick, and Canadian landscapes framed on liquor billboards, and the volume of diffusion takes on substantial proportions.

In sports, Canada's obvious contribution is ice hockey. With National Hockey League expansion during the 1970s, large numbers of Canadians came to the United States to play in cities as far away as Los Angeles and St. Louis. A more subtle but lasting impact results from the young Canadian players recruited to U.S. universities where they interact with students and faculty as well as play their game. At the University of Maine, where hockey is a major varsity sport, Canadian players contribute substantially to attitude adjustment, interest, and appreciation of Canada.

Close scrutiny of the American business scene reveals a distinct Canadian presence. Of interest in this consideration of cultural transfers is the Canadian way of doing business, parallel to the Japanese assault on the U.S. market. It is proving highly successful in banking and real estate sectors where large Canadian chartered banks with massive assets and nationwide real estate concerns enter the United States with highly developed organization and understated finesse. In major cities and development areas like the Sun Belt, Canadian firms are buying a share of U.S. growth and recovery.[15] Leading the incursion are companies like Cadillac Investments with an American name and an origin in the years of U.S. business domination in Canada. In effect, a multinational turnaround is occurring with Canadian firms like Northern Telecom competing effectively with U.S. communications companies in the United States and abroad. Canada's success in the communications field is further illustrated by the ample dissemination of productions from the Canadian Broadcasting Corporation to educational and public television facilities throughout the United States. Prominent among these are children's shows like the currently popular "Fraggle Rock" originating in Toronto.[16] Like the "Canadian Arm" on the U.S. space shuttle, they proclaim a growing Canadian presence in North American business and communications.

In the borderland, transfers are readily visible and often more apparent than Canadian leapfrog efforts to the Sun Belt or hierarchical diffusion through the urban network in the United States. The sunflower landscape of southern Manitoba centered on Altona has spilled over the border and now characterizes tracts of North Dakota farm-land as well. In the Pacific northwest distinctive British Columbia housing styles reminiscent of British cottages are found in Bellingham, Seattle, and other Puget Sound cities.

The Madawaska Twin Barn as an Indicator of Cultural Adjustment in the Borderland

A detailed example of material culture transfer from northern Maine serves to illustrate the dynamic of borderland adjustment in historic context. French Canadian migration to New England during the nine-teenth and twentieth centuries was usually a rural to urban process in which immigrants left material culture and settlement characteristics behind and brought with them portable culture-language, religion, and social institutions. In the St. John Valley of Maine and New Brunswick, Acadian and Québecois settlers established a distinct cultural enclave prior to designation of the international boundary. Acadian refugees arrived in the valley late in the eighteenth century and were joined by Québecois from the south shore of the St. Lawrence throughout the nineteenth century.[17] Together they formed the République du Madawaska along the upper St. John River valley, an isolated agricultural settlement characterized by distinctive furniture, textiles, land use, architecture, and even woodpiles.[18] In 1842 the Webster-Ashburton Treaty between Britain and the United States es-tablished the boundary along the river and through the heart of the culture area, but the border brought few changes until the "potato boom" of the late nineteenth century began differentiating Canadian and American settlements.[19]

An isolated setting and a well-recorded sequence of events provide an opportunity to explore fundamental questions of how culture is transferred across a national boundary, at what rate, by whom, and in what form. From the time of designation the boundary sometimes acted to halt exchange of ideas and practices and at other times to filter them. The back-and-forth flow of intrinsic components of French cul-ture — language, family structure, and religion — were only slightly affected by the border. At first this appears true of the substantial legacy of barns, but detailed study shows striking differences and pro-vides a sensitive record of culture transfer.

As respositories of culture, few artifacts rival barns, for they are sub-stantial and conservative structures which record clues to settlers' origins, the routes they took, and eras when they arrived. Faithful replication of traditional barn features by settlers often provides outlines of culture areas[20] and strong indications of cultural identity in place.[21] In this sense barns are illustrative of local variation and personal preference while sustaining or adjusting cultural tenets. Owing to the cost of removing obsolete barns, several generations of structures may coexist in the same farmyard to document the sequence of barn building practices. These characteristics, added to the relative ease with which barn features may be recorded and their ubiquity on the land, establish them among the most useful farm artifacts for evaluating culture change in an area where agriculture predominates.[22]

In the St. John Valley of Maine and New Brunswick, four barn types prevail: Acadian, Quebec, Madawaska twin, and New England connecting (Fig. 1). Characterized by a side entry and simple square plan, the Acadian barn has a central threshing floor, hay mow on one side, and granary and stables on the other. Later forms exhibit a dust-pan extension for additional stable space and exterior shingling for insulation of the stable area. Quebec barns maintain the tradition of aligning several threshing floors with side entries in a rectangular structure with upswept eaves (Norman roof), a few small windows distributed irregularly about the barn, and distinctive geometric door decoration. Inside, the stables are at one end with the hay mow above extending the length of the barn. The Madawaska twin barn consists of two identical sections in parallel alignment and connected by a passageway (*tambour*) which ranges from one story to roof-ridge height. In one section are livestock stalls, hay mow, and toolshed, and in the other are spaces for threshing, granary, machinery storage, and hay. Early twin barns are gable-roofed structures with little or no shingling, whereas structures built in the 1920s and 1930s are usually shingled with distinct gambrel roofs or gable roofs with connected ridges. New England connecting barns appear in both gable-end and side-entry versions constructed of vertical boards, clapboarded and painted white. Interior arrangements vary considerably although hay mow, threshing floor, and granary are usually located in a large structure and stables for horses in an adjoining section. Both are connected at right angles or in tandem to the woodshed and house.

Among these barn types, the Madawaska twin incorporates certain features of Acadian and Québecois antecedents and bears an affinity to New England connecting barns. These structures are connected to each other but never to outbuildings or the house. Found almost exclusively

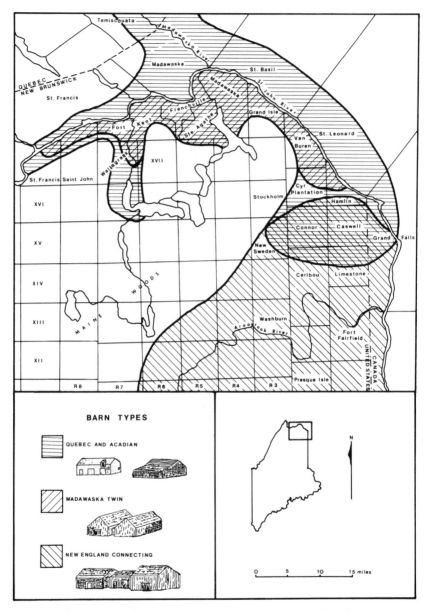

Fig. 1. The distribution of barn types in the northeast border country. Based on surveys by Victor Konrad and Michael Chaney, 1980-1982. Map reprinted from the *Journal of Cultural Geography* 3:1 (Fall/Winter, 1982), p. 71. Reprinted with permission of Bowling Green University Popular Press.

in the French-inhabited valley between St. Francis and Hamlin, Maine, the Madawaska twin barn represents an alternative to the New England connecting barn which prevails in the rest of agricultural Aroostook County.[23]

Valley craftsman Israel Ouellette of Ste. Agathe, the Bechard brothers and Alphonese Pelletier of Van Buren, and Victor Morin of Grand Isle planned and guided construction, but neither they nor master barn builders Will Beckwith and Cluny McPherson, Sr., who worked throughout the county, originated the twin barn concept. According to informants the idea originated in Quebec and was brought to the St. John Valley by immigrants like Francois Faucher who came from the only area in the province where the twin barn was built.[24] *Les granges jumelées* are found only south of the St. Lawrence River between Sorel and Drummondville, in Drummond and Bagot counties.[25]

Most Quebec twin barns predate the construction of Madawaska twin structures; similarities in design, layout, and materials are striking and further confirm the Quebec origin.

In the St. John Valley, twin barns appeared in the late nineteenth century in gable roof form, then evolved through hybrid gable/gambrel form to separate lines of gable and gambrel twins (Table 2). Evidence suggests that hybrids and some other twin barns were accomplished by moving a barn to another location beside an existing barn and then connecting the two.[26] Alternately, the practice was to build one barn, construct the other beside it when needed and then connect the two. Some of the structures were joined in an attempt to conform to the practice of twinning barns; that is, some twin barns do not reflect the evolution of the practice but rather provide evidence of its acceptance (Table 2). Valley farmers often rejected the gambrel roof form because it marked an American barn, whereas the gable roof was considered more French Canadian. They claimed reluctance to accept the gambrel roof for the practical justification that gable roofs shed snow more easily and gambrel roofs eventually failed under the strain.[27] There is no evidence to support this contention.

The concentrated and limited twin barn distribution is restricted to the river communities and adjacent agricultural expansion townships of Ste. Agathe, Wallagrass, and Cyr Plantation, all on the United States side of the border (Fig. 1). Acadian and Québecois barns are found throughout and beyond the twin barn distribution, whereas the New England connecting barn extends only to Presque Isle and Caribou, the border towns of English-speaking Aroostook. Because of

TABLE 2. Madawaska Twin Barns Classified by Decade of Construction and Type of Roof

| Decade of Construction* | Type of Roof | | | | | | Twin Barns Dated | Percentage Twin Barns Dated |
| | Gable | | Gable/Gambrel | | Gambrel | | | |
	No.	(%)	No.	(%)	No.	(%)		
1881-1890	1	(100)					1	2
1891-1900	5	(83)			1	(17)	6	14
1901-1910	7	(70)	2	(20)	1	(10)	10	23
1911-1920	4	(44)	1	(12)	4	(44)	9	21
1921-1930	5	(42)			7	(58)	12	28
1931-1940	3	(60)			2	(40)	5	12

*Based on verified dates marked on barns and on information provided by owners and other informants interviewed.

Source: Victor Konrad and Michael Chaney, "Madawaska Twin Barn," *Journal of Cultural Geography* 3 (Fall, Winter 1982): 69.

French settlement expansion, some Acadian and Québecois barns are found in the New England-connecting-barn area, but Madawaska twin and New England connecting barn distributions do not intersect. Within the twin barn distribution area concentrations are greatest in agricultural expansion areas farthest from the river and immediately adjacent to the St. John Valley where twin forms replaced earlier deteriorated barns.

The Madawaska twin barn was developed as a distinctly French adaptation to expanding agricultural requirements in a potato growing area. In adjacent New Brunswick, predominantly French as well, growing potatoes or other crops on a large scale was not important during this era. In the predominantly English-speaking Aroostook Valley, the need for more barn space saw the construction of larger connecting barns or single structures.[28] Originating in Quebec, the twin barn concept was adapted in Maine to symbolize French Canadian tradition when the move to a new and larger barn was required. It fuses stong and distinctive elements of the two cultures by combining the idea of connecting barns with the juxtaposed tradition of not connecting the barn and the house. The twin barn allows advances in form and construction required by agricultural change and consistent with majority Yankee culture, but it sustains the unmistakable tie to French Canadian barn-building tradition. Distribution of the Madawaska twin barn marks the boundary of a restricted but viable culture area where differences between French Canadian and American culture were not removed but redefined in the process of cultural adjustment.

Ebb and flow of Canadian culture continues in the upper St. John Valley of Maine, for in the last decade reawakening of French Canadian identity (coupled with and perhaps spurred by Quebec nationalism and Acadian resurgence) is becoming increasingly visible in valley landscapes. The Norman roof, hallmark of contemporary Quebec housing, is found on several new structures between Fort Kent and Madawaska. New barns, often no more than elaborate two-car garages, sport second-story hay doors and main door decoration reminiscent of historic Quebec barns. These are tangible examples of Canadian culture extending into the United States on a recurring basis. Sometimes retained and sometimes repudiated, material culture elements are usually altered or adapted to meet special requirements in the region. Like the Madawaska twin barn they should be interpreted as markers of cultural persistence and resurgence as well as hybrids of material culture transition. The complexity of culture transfer in the

St. John Valley cautions against viewing the international boundary between Canada and the United States as a simple line across which culture either does or does not extend, or a border which is consistent from coast to coast.

Cross-Border Transfer and the Development of Borderland Cultures

Regional distinctiveness of cross-border cultural transfers is emphasized in the Maine barn study as well as in previous illustrations. Along the extensive Canada-United States boundary, the distinct nature of cultural adjustment by region and often locale, stands in direct contradition to the federally imposed management of a consistent boundary. Segments of the boundary are different, reflect varying cultural transfer characteristics, and require adjusted policies to accommodate these differences. In Canada the policy of recruiting both immigration and customs officials from local border areas helps to assure greater appreciation of local and regional conditions and cross-border problems. But in both Canada and the United States regulations as well require adjustment to local and regional conditions. In some instances local organizations like the St. Stephen (New Brunswick) and Calais (Maine) Chamber of Commerce are well integrated to promote cultural adjustment as well as business across the border. At many points along the border, international commissions enhance operation of national parks (Glacier-Waterton Lakes), tourist areas (Niagara Falls), water resources (Columbia River Project), transportation (St. Lawrence Seaway Authority), and cultural resources (Roosevelt-Campobello International Park).

Also conveyed in the preceding discussion are three additional insights about cultural transfer from Canada to the United States. First, the movement of Canadian cultural elements across the border is substantial in sheer magnitude as well as in variety of elements. The barrage of American culture to Canada occasions a substantial response, and Canadians are indeed beginning to meet the challenge despite the overpowering presence of the United States. Second, Canadian cultural transfers to the United States are subtle. In the face of the overwhelming U.S. presence, understated advances appear more effective than blatant challenges. And third, the transfer of culture is constantly changing in response to both Canadian and American cultural transportation. This dynamic, this constant adjustment, sustains the cultural transfer process and keeps it vigorous.

These characteristics, coupled with regional distinctiveness, describe an evolving borderland culture zone charged with energy and change. It is inaccurate to characterize this as a consistent transition zone, another region, an outer "sphere" of culture or a "zone of penetration,"[29] for it contains dynamics at national, regional, and local levels which are not accounted for in these concepts. In addition, Canada's borderland with the United States contains the Canadian heartland insuring a greater coincidence of national policy and borderland requirements than is possible in either the United States or Mexico. The borderland is more significant to Canada than to the United States. Canada's cultural identity depends on a strong presence and renewed culture transfers into the border area. Consistent with the model of accommodation, these transfers aim to adjust cultural differences between Canada and the United States and result in coalescent features on the landscape in between. Since the border is conveyed as a linear geopolitical concept, most Canadians and Americans labor under an image of abrupt differences between the two countries when in fact a borderland of considerable interaction exists. To see this landscape requires recognition of the cultural transfers and an acceptance that they are indeed benign events in the accommodation of two distinct but compatible cultures.

Notes

1. John Porter, *The Vertical Mosaic* (Toronto: University of Toronto Press, 1965).

2. Wilbur Zelinsky, "North America's Vernacular Regions," *Annals, Association of American Geographers* 70 (March 1980): 1-16.

3. Andrew P. Vayda, *War in Ecological Perspective, Persistence, Change, and Adaptive Processes in Three Oceanic Societies* (New York: Plenum, 1976), 1-9.

4. Marshall Sahlins, *Historical Metaphors and Mythical Realities, Structure in the Early History of the Sandwich Island Kingdom* (Ann Arbor: University of Michigan Press, 1981), 33-66.

5. Mel Watkins, "The Dismal State of Economics in Canada," in Jan Lumsden, ed., *Close to the 49th Parallel, Etc.* (Toronto: University of Toronto Press, 1970), 197-208.

6. L.D. McCann, "Heartland and Hinterland: A Framework for Regional Analysis," in L.D. McCann, ed., *A Geography of Canada, Heartland and Hinterland* (Scarborough, Ontario: Prentice-Hall, 1982), 3-35.

7. Wilbur Zelinsky, *The Cultural Geography of the United States* (Englewood Cliffs: Prentice-Hall, 1973), 113.

8. Deryck Holdsworth, "Regional Distinctiveness in an Industrial Age: Some California Influences on British Columbia Housing," in Victor Konrad, ed., "The Transfer of Culture on the Land Between Canada and the United States," *The American Review of Canadian Studies* 12 (Summer 1984): 64-81.

9. James Marston Fitch, *Historic Preservation* (New York: McGraw-Hill, 1982).

10. Kevin Lynch, *What Time is This Place?* (Cambridge, Mass.: The MIT Press, 1972).

11. Peter Ennals, "The Yankee Origins of Bluenose Vernacular Architecture," in Konrad, ed., "The Transfer of Culture," 5-21.

12. R. Cole Harris, "Regionalism and the Canadian Archipelago," in McCann, ed., *A Geography of Canada*.

13. Eric Waddell, "Cultural Hearth, Continental Diaspora: The Place of Quebec in North America," in McCann, ed., *A Geography of Canada*, 147-48.

14. Val Ross, "The Joys of a Bountiful Season," *Maclean's* (19 December 1983): 40-48.

15. Fred Lazar, *The New Protectionism: Non-Tariff Barriers and Their Effects on Canada* (Toronto: Lorimer, 1981); Grant L. Reuber, *Canada's Political Economy: Current Issues* (Toronto: McGraw-Hill Ryerson, 1980).

16. Patricia Hlucky, "New Programs for the Thinking Child," *Maclean's* (9 January 1984): 40-42.

17. James Allen, "Franco-Americans in Maine: A Geographical Perspective," *Acadiensis* 3 (Autumn 1974): 32-66.

18. Victor Konrad, "From French Canadian to Franco-American: Late Nineteenth Century Settlement Change in the Upper St. John Valley, Maine," *Proceedings, New England-St. Lawrence Valley Geographical Society* 10 (1980): 15-22.

19. Victor Konrad, "Against the Tide: French Canadian Barn Building Traditions in the St. John Valley of Maine," in Konrad, ed., "The Transfer of Culture," 21.

20. Wilbur Zelinsky, "The New England Connecting Barn," *Geographical Review* 48 (October 1958): 540-53.

21. Henry Glassie, *Pattern in the Material Folk Culture of the Eastern United States* (Philadelphia: University of Pennsylvania Press, 1968).

22. Konrad, "Against the Tide," 22-23.

23. Victor Konrad and Michael Chaney, "Madawaska Twin Barn," *Journal of Cultural Geography* 3 (Fall, Winter 1982): 64-75.

24. Michael Chaney, "Journal of Fieldwork, St. John Valley, March 2-5, 1981, Twin Barns Interviews," University of Maine, Orono, Northeast Archives of Folklore and Oral History, 1464 (1981): 104-153.

25. Robert-Lionel Seguin, *Les Granges du Quebec du XVII au XIX siècle* (Ottawa: Musée National du Canada, Bulletin 192, 1963): 81-82.

26. Chaney, "Journal of Fieldwork," 116, 139-43.

27. Ibid., 143, 146, 149.

28. Cluny McPherson, Sr., "Aroostook Barns," in *Faces of Aroostook, A Bicentennial Portrait of Aroostook County* (Presque Isle, Maine: Polar Star Associates, 1976), 32-33.

29. Donald W. Meinig, "The Mormon Culture Region: Strategies and Patterns in the Geography of the American West, 1847-1964," *Annals, Association of American Geographers* 55 (June 1965): 213-17; Donald W. Meinig, *Imperial Texas: An Interpretive Essay in Cultural Geography* (Austin: University of Texas Press, 1969), 110-24.

Part V

African Borders

Problem Solving Along African Borders: The Nigeria-Benin Case Since 1889[1]

A. I. Asiwaju

Introduction

The problems posed in international boundaries are essentially those of human relations; since problems are a function of perception, it is necessary to stress the three main levels involved in border relations. First, the formal, which directly involves the interacting national states and governments, is the official level, the domain in which diplomacy and international law operate as sole instruments of control and conflict resolution.[2]

Next, in descending order, biased in favor of the official mind which creates borders in the first place, is the level characterized by relations between local governments or administrations interacting across borders. This second level falls short of the official or formal status only because of the absence of sanctions by the sovereign governments concerned. This absence renders border relations at this level fragile and vulnerable to disruption through possible interference of territorial authorities. The instrument of control operates on the basis of informal agreements and interpersonal understanding among local government officials. Ivo Duchacek, in a reflection on the practice of relations on the U.S.-Mexico border, has labeled this strategy "subnational microdiplomacy."[3] Although this second area (so well focused in studies of the U.S.-Mexico and Western European situations) has been described under the category of "informal linkages" in the existing literature,[4] it will be referred to in this paper as the semiofficial.

We reserve the category of "informal" for the third and the more completely informal level where neither the territorial governments nor the officially recognized local authorities play any appreciable role. This is the level of the local indigenous communities. Relations at this level show up clearest in situations where nation-state boundaries are superimposed on frontiers of preexisting human groups, culture areas or traditional states. Borders are often drawn across such units causing the artificial division of the coherent entities. Under such a condition, represented across the globe by the Allemanic subgroup of the Germans, is the area of the "regio" fragmented by the Franco-Swiss-German borders; the Catalans and the Basques astride the Franco-Iberian frontier; the Chicano-Mexicano of the U.S.-Mexico border; the Iroquois Indians of the Canada-U.S. boundary; and the Kazakh across the Sino-Soviet frontier. Border relations result from the operation of age-old, intragroup relations on the model of what Linda Whiteford has rather appropriately described as "an extended community."[5] The essentially humanistic concern of border relations at this level is demonstrated in the emphasis on communal feeling or group consciousness based on cultural and kinship ties or traditional state loyalties.

The point about these different levels of operation is that they strongly influence, if not determine, the perception of border problems. The question as to what is or what is not a problem often has to depend not only on the interests and priorities of each of the interacting national states but also on the specific concerns at the distinct levels of the local governments or administrations and the extended communities. Each of these dictates its own perspective. Border problems arise to the extent of conflicts within the network of relations at one particular level (usually that of the interacting national states) as well as between the levels (usually between the first and each of the other two lower levels). The main challenge to policy is in establishing the necessary harmony within and between the levels. Solutions proposed would be appropriate to the extent to which they help in reducing, if not altogether bridging, the gaps in perceptions and in reconciling the perspectives.

Discernible trends in border research suggest a certain amount of regional emphasis and periodic variation in the distribution or occurrence of the three categories. Thus, while Western Europe, with particular reference to the Rhineland, has been fully exposed to all three levels and is currently demonstrating to the rest of the world the possibility for harmony within and between the levels,[6] studies of relations across the U.S.-Mexico border have emphasized more the formal and semiformal levels with little or no systematic research on the informal linkages as discussed in this paper.[7] The African experience is of special interest because of the prominence of the formal and the informal levels of border relations and the peculiarity of the stunted growth of the semiformal category.

While this paper will strive to offer brief discussions of all three levels of operation as they manifest themselves on the Nigerian-Benin border in the period indicated, it will stress the informal level of the African experience because of the opportunity it offers for placing emphasis on the humanistic dimensions. The humanistic dimensions are also discernible in Europe and North America.[8] However, they have been far less visible and, in terms of systematic research, less focused in the highly industrialized and urbanized borderlands than in Africa (and perhaps Asia) where frontier zones are characteristically areas of largely rural and village-level societies and closely knit kinship groups.

The Nigerian-Benin Border

The choice of this African case study is based on several considerations. These include a demonstrable representative character,[9] availability of a reasonably large body of researched material, the author's more intimate personal knowledge and direct research experience[10] and, above all, the quite appreciable extent of comparability of structure and functions vis-à-vis the U.S.-Mexico border.[11]

As several studies have shown,[12] the Nigerian-Benin border shares with other African boundaries the history of European imposition. It was created as a consequence of the international rivalry between Britain and France, the colonial powers dominating the history of the European scramble for the partition of Africa in the last quarter of the nineteenth century. As elsewhere in the continent, particularly West Africa where most borderlands are of Anglo-French configuration, the Nigerian-Benin neighboring border regions represent areas of distinct official languages and national histories and cultures, as well as differing economic, legal, political, and administrative systems and institutions. Like other borders, including those in Africa, the one between Nigeria and Benin has been justly described as artificial. While Anene[13] has sought to demonstrate the rational basis for its definition, allocation, and final demarcation, the frontierline is known to have split coherent culture areas and distinct economic regions. Although border disputes and conflicts have been relatively few, the incursion of Benino

gendarmes into Illo areas of the Sokoto State of Nigeria in 1981 has
sufficiently indicated the potentials for such conflict.

The comparison with the U.S.-Mexico situation has elsewhere been
discussed on a relatively larger scale;[14] it should suffice, therefore, to
draw attention to the main features as summarized in the Appendix. It
follows from this comparison of the structures and functions of the two
border situations that relations on the Nigeria-Benin boundary cannot
be and indeed, are not fundamentally different from what they have
been on the U.S.-Mexico border.

Without disregarding obvious local particularities and regional
variations in the scale of the operations, one can argue that the basic
problems relating to unauthorized movements of people and materials
are substantially the same on the two sets of borders. Also comparable
are issues of interlocking interests and resources and the semiformal
and informal linkages which (a) function to neutralize local effects of
conflicting mandates of the interacting national states and (b) provide
appropriate solutions to the local problems posed by the border. The
semiformal and informal networks in both the Nigeria-Benin and
U.S.-Mexico borderlands have derived strength from the local inhabitants who commonly view their respective national governments
as insensitive to problems created for them by the presence of the
border in their midst. Nigeria, as the "Giant of Africa," relates to Benin
in much the same way as the U.S., the most super of the world "super
powers," relates to Mexico. Consequently, the relations between
Nigeria and Benin, under the former name of Dahomey, in the 1960s
were at times as influenced by "fears of aggression" on the part of
Dahomey[15] as has been the case of Mexico vis-à-vis the U.S.

As between the U.S. and Mexico, basic disparities of national interests and priorities exist between Nigeria and Benin. There are discernible gaps in perception even of the same border issues. Studies of
contraband trade across the Nigeria-Benin boundary point to such important instances, as for example, the massive smuggling of Nigerian
cocoa from 1970 to 1975,[16] when activities outlawed in one state
became legitimate business in the other state across the border.

As we shall presently see, this disparity of perception — inspired
partly by the contrasting size of the territories, populations, and
economies of a Nigerian Goliath vis-à-vis a Dohomeyan David and
partly by the differing political cultures of the two states — has constituted a major obstacle to genuine cooperation between the national

governments in spite of an early realization of the need for collaboration and some efforts to achieve it.

History of the Border

The Nigeria-Benin border is exactly the same Anglo-French colonial boundary that demarcated the area of what used to be the British Colony and Protectorate of Nigeria to the east and the French West African colony of Dahomey to the west. Negotiated in two separate sections — the southern from the Atlantic coast to latitude 9° north, based on the Anglo-French Agreement of 10 August 1889, and the remainder to the north, based on the Anglo-French Protocol of 14 June 1898 — the boundary was the direct result of the international rivalry between Britain and France beginning on the Atlantic coast in the early 1860s.[17] Actual demarcation took place as a result of the work of a series of Anglo-French Boundary Commissions in 1895-96, 1906, and 1912, when boundary markers were erected and reerected along the binational line;[18] minor diplomatic adjustments took place in 1927, 1937, 1952, and 1960,[19] but the main alignment has remained as defined in the original agreements of 1889 and 1898 and comprehensively set out in the Agreement of 14 October 1906.[20]

In 1960, following the attainment of sovereign status by French Dahomey in August and British Nigeria in October, the boundary shed its colonial reference and became the Nigeria-Dahomey international boundary. The former French colony, which became the Republic of Dahomey in 1960, underwent a second rebaptism in 1975 when it became the République Populaire du Benin (The Peoples Republic of Benin). This occurred in the wake of an officially proclaimed socialist revolution following the last in the series of political crises which had plagued the young nation-state since 1963, making it the independent African state with a yet unbeatable record of coups d'états. With this change, the Anglo-French colonial border, which became the Nigeria-Dahomey international boundary in 1960, took its present appellation as the Nigeria-Benin border.

The border compares with most other international boundaries in terms of its characteristics as a line of demarcation between two inequalities. We have already pointed to this fact as it applies to the contrasting sizes of territory and population and to the scales of the two economies. The point is also demonstrated in the mutually antagonistic

Anglo and Latino cultures which dominated the colonial and post-colonial governments and officials on the respective sides of the borders. Officially, Nigeria was and is still an Anglophone country while Benin has remained Francophone. Each operates political, legal, and administrative traditions and institutions inspired by the two distinct European cultures. In Nigeria, for example, the legal tradition is on the British model and the associated principles of the Common Law, whereas in Benin and other Francophone African countries, the tradition has been based on the civil law practice on the Roman model.[21] Nigeria was administered as a federated colony and is still federal in government and decentralized in administrative tradition. Benin has operated strictly as a centralist state; it was administered in the colonial era as a unit in the French West African colonial grouping based in Dakar. Given the similarities of the state structures between Nigeria and Benin on the one hand and, on the other, the U.S. and Mexico,[22] the official culture contact situations on the respective binational boundaries are bound to be closely related.

The Border Impact

To appreciate in full the partitioning effect of the Nigeria-Benin border on the indigenous cultural landscape, a good knowledge of the culture areas prior to the partition is necessary. Three concentric cultural ecumenes were situated in areas through which the border was drawn. First were the specific communities of identical culture, kinship ties, and political organization, which lay directly on the path of the border when it came to be drawn. The southernmost of these were the Gun subgroup of the Aja-speaking peoples in Badagry (Nigeria) and Porto Novo (Benin) regions.

Then followed the six western Yoruba subgroups of the *Awori* in Ado-Odo (Nigeria) and Itakete (Sakete on French maps) in Benin; the *Anago* in Ipokia, Ijofin, Ibatefi, and Agosasa (Nigeria) and Porto Novo (Benin); the *Ifonyin* in Ifonyintedo, Ilase, Ihumbo, and Idologu and Abalabi (Nigeria) and Ifonyinle (Benin); the *Ohori* (Holli or Hollidge on French maps) in Ije, Ipobe (Pobe in French documentation) Isale, Ilemon, Iganna in Benin and Orisada, Ibayun, Imoto, Ikotun and Ohumbe in Nigeria; the *Ketu* concentrated in Imeko, Idofa, Ilara, Ijale, Ijoun, Igan Alade, Egua and Ebute Igboro in Nigeria, and Ketu, Ofia, Dinrin, Iselu, Ilikimu in Benin; and finally, the *Sabe* in Iganna, Oke-Iho, Sabe (mid-way between Iseyin and Saki in Dyo State) and

Iwoye (Nigeria) and Ile-Sabe (Save in French documents), Kokoro, Kabua, Kilibo, Sala-Ogoi, Saworo (Tchaourou) and Jabata in Benin. North of the Western Yoruba subgroups was the Borgu Kingdom of Nikki, split into two by the border so that Nikki (the capital) was placed in Benin while the rest of the traditional state, including Yashikera, Kaima, Okuta and Ilesa, was situated on what became the Nigeria side of the border.

These specific communities of the Aja, the Yoruba, and the Borgu together constitute the area most directly impacted by the border. The sample settlements enumerated for each on the opposite sides of the border are as intricately interwoven, if not more intimately so, as the more researched twin cities of the U.S.-Mexico border or their European counterparts in the Regio and across the Rhine River. Badagry and Porto Novo, Imeko and Ketu, Yashikera and Nikki, though not so directly located on the border and not as vast or sophisticated, are as socially and economically interlocked as El Paso and Ciudad Juárez and Basel and Mulhouse.

One observable difference of emphasis distinguishes between the African and the Euro-American situation. In the Nigeria-Benin case, as with other known African border ethnographic situations (including those of the Somali fragmented by the Ethiopia/Somalia/Kenya borders, the Il Maasai astride the Kenya-Tanzania boundary, specific Bakongo communities across Gabon/Congo/Zaire/Angola frontierlines, to mention a few), the networks are cultural and social before they are economic. To be sure, considerable material motivations and classifiably economic activities operate most especially in the area of trade and mobility of labor of both rural-rural and rural-urban orientations. Such activities may be legitimate or not, depending on the angle and the level of perception; but, more often than not, they take place in the context and in the general service of impacted kinship groups and in consideration of the survival and overall welfare of the communities.

The specific communities directly affected by the border were linked to the wider areas of each of the three main culture areas of the Aja, the Yoruba, and the Borgu. Thus, the Gun, who are directly affected by the boundary arrangement, were linguistically, culturally, and historically linked to the other Aja-speaking peoples to their east and more importantly to their northwest and west, including the Wemenou of the Weme River Valley and the Fon of Agbome (Abomey), Allada, and Whydah.[23]

Similarly, the Western Yoruba subgroups astride the border share a sense of community with the rest of the Yoruba culture area in present-day Lagos, Ogun, Oyo, Kwara, and Ondo States of Nigeria to the east,[24] and the more westerly subgroups such as the Jaluku, Mayinbiri, Ana, and Fe in the middle latitudes of Benin and Togo republics.[25] Nikki, whose specific territory was sliced into two by the Anglo-French border, used the same language, traditions of origin, political, and social institutions, and overall group consciousness with all other Borgu states centered on Bussa and Illo in the Kwara and Sokoto states of Nigeria and Kandi (Kouande) west of Nikki in Benin.[26]

While the three culture areas remain distinct in terms of language and specific geographical location, all are historically linked and culturally interconnected. Several concrete studies have supported the easily observed phenomenon of the cultural and historical affinities between the Yoruba and their Aja and Borgu neighbors. Such affinities have been found to be so numerous and complex that the frontiers between these three ethnic groups are extremely difficult to place.[27] Individually and severally, the three culture areas were also known to have operated within a much wider cultural ecumene which has been argued to include Nupeland to the north, the Edo-speaking peoples in the area of present-day Bendel State of Nigeria to the east, and the Ewe of present-day southern Togo and Ghana.[28]

This picture of ethnic coherence and interlink is important for an understanding of the evolution of the border under discussion. First, the border exercised a very important influence on the pattern of the Anglo-French international rivalry in the Gulf of Guinea, which led ultimately to the partitioning of the area between the two powers at the end of the nineteenth century. European interest in this part of Africa, as elsewhere in the continent, was predominantly economic with particular reference to trade. European expansion into Aja-Yoruba-Borgu parts of the West African interior was a direct response to the realization of this whole area as the vital hinterland for British and French trading activities in Lagos and Porto Novo.

The question of economic and social interlock between Lagos and Porto Novo, and between them and the Aja-Yoruba-Borgu interior, has continued to militate against the barrier functions of the Nigeria-Benin border. In terms of its outermost circle, the overall cultural ecumene on which the Anglo-French boundary was superimposed demonstrates that the impacted area is remarkably more extensive than the immediate neighborhood of the border.[29] Scholarly studies and policy analyses concerning topical border questions such as smuggling and "alien invasion" will remain inadequate so long as they continue to place emphasis on the strictly economic factors to the exclusion of the human dimensions indicated in the intricate network of intra- and interethnic relations across the border.

Binational Relations Under Colonial Rule[30]

In the period from 1889 to 1960, binational relations with regard to the border between Nigeria and Benin (then Dahomey) were, as with all other Anglo-French colonial boundaries in West Africa (Nigeria's other four borders inclusive), no more than the relations between the colonial territorial governments and department local administrations set up in the two respective colonial possessions by Britain and France. For reasons of the parallel characters of the two metropolitan cultures and the rivalry between them as colonial powers in Africa, the two administrations operated essentially as competitive rather than as complementary entities.

Thus even when they recognized obvious difficulties in keeping the two governments apart, the two administrations ultimately had to settle for arrangements based on mutually exclusive and particularistic interests. For instance, instead of the provisions in relevant bilateral border agreements for freedom of trade by European nationals in all colonial possessions, irrespective of which of the two powers was in control, or those guaranteeing cross-border movement for members of the local indigenous communities, the rival British and French colonial administrations settled for developments in the direction of preferential tariffs and border control. With particular reference to the Nigerian border in question, the erecting of boundary pillars in 1906 and 1912 and the establishment between 1900 and 1942 of customs posts by authorities on both sides[31] were steps taken to transform the boundary into an effective political, as well as economic, dividing line.

Yet the illogic and artificiality of the new boundary continued to manifest itself. The two colonial administrations were often forced by local realities to abandon considerations of national pride and mutual isolation to communicate with each other in order to cope with several common problems and interests. First was the impact of the strong historical, ethnological, cultural, and kinship ties between subject peoples on both sides of the new intercolonial boundary. This point, so easily proved by the network of relations within and among the Aja,

Yoruba, and Borgu along the Nigeria-Benin border, has been of general applicability to the wider area of Africa.[32] With particular reference to the West African subregion, characterized by the phenomenon of Anglo-French colonial border features, other specific culture areas also split into British and French spheres have included those of the Ewe of present-day Ghana and Togo; the Hausa of Northern Nigeria and Southern Niger; the Wolof and the Serrers of Senegal and the Gambia; the Kanuri and Shuwa Arabs astride Nigeria-Chad-Cameroon borders; the Mandara, the Fulani, the Mingi, the Chamba, the Ododop, the Efik across the remainder of Nigeria-Cameroon border south of Lake Chad, and the Gourmantche who straddle the Benin-Togo-Upper Volta-Niger borders.

In all these and similar situations, the inevitable movement of people across the intercolonial boundaries naturally reduced the effectiveness of the boundaries as dividing lines. Especially has this been the case with respect to situations like those of the specifically partitioned Aja, Yoruba, and Borgu communities, where divisions were not just those of cultural entities but also those of closely knit kinship groups and economic units including farmlands, fishing rights in riverain areas, and trade.

Secondly, political and administrative problems were of mutual interest to the different colonial administrations. Of particular significance was the problem of political agitation by subject peoples, whether in the form of armed revolts or protest migrations. Where such unrest related to peoples strategically situated along or close to intercolonial boundaries, normal police action on the part of the affected colonial authority was often obstructed by the need for diplomatic considerations for the neighboring power. Under such a condition, effective management of the crises always called for cooperation and some measure of joint action of the interacting territorial authorities.

Politically discontented West African peoples along borders displayed their resourcefulness by turning into their advantage the protocols which regulated the behavior of European governments along interstate boundaries. Ring leaders of abortive revolts often headed towards the intercolonial boundaries with many often escaping the offended power's police dragnet by crossing over into the area of jurisdiction of a different but neighboring colonial authority. Thus, for example, a number of the ring leaders in the Iseyin-Oke-Iho Rising of 1916 escaped from the area of British jurisdiction in Nigeria by crossing into French Dahomey.[33]

In some other cases of armed revolt, such as frequently took place among the Ohori of Ije in French Dahomey during the colonial period, a rebellious situation along an intercolonial state boundary constituted a very serious embarrassment to the French authorities, since the Ohori, because of their relations across the border with Nigeria, had access to a fresh supply of arms. Ring leaders often found it easy to cross over to the British side where their kinsmen actively cooperated and frequently succeeded in hiding them away from the British authorities, who often made half-hearted attempts to make arrests in cooperation with the French.[34] Similar events have been recorded with respect to the people of Borgu when they revolted at different times against both their French and British rulers.[35]

More spectacular and certainly more serious than the issue of armed revolts was the question of protest migrations, which constituted a major factor in Anglo-French diplomatic communications in West Africa during and in-between the two World Wars. Protest migration, most often from French to British spheres, was a common experience all over West Africa where French rule proved decisively more strenuous and a lot more demanding than the British. Emigration away from the area of jurisdiction of a resented political authority was an old and universally known form of revolt. In former French West Africa, as elsewhere in the era of European colonial rule, protest migration eventually gained an ever-widening acceptance, probably because of the subject African peoples' awareness of the futility of confrontation with the usually superior arms and military strategy of the white man.[36]

Examined in terms both of the "push" and "pull" factors, the protest migrations have involved considerations of differences as between the departure and arrival ends (in this case the French and British colonies respectively) in matters of civil obligations such as conscription, forced labor, taxation, and requisition. Other considerations "forcing" people out of French West Africa included the suppression of traditional political institutions and excessive official control. Apart from the West African Aja, Yoruba, and the Borgu in French Dahomey; the Hausa in Niger; the Sanwi, the Affemas, the Baoule, the Agni, the Abron, the Kulango, the Lobi, and the Dagari in the Ivory Coast; the Mossi and the Dyula in the Upper Volta; and the Wolof in the Sine-Saloum region of Senegal were all among such peoples who for more or less similar reasons fled French territories and took refuge in the various neighboring British colonies.[37] The fact that each of these groups had been split into two at the time of the Anglo-French partition meant that the migrants always had their kinsmen in the British localities where they fled and which facilitated their movements and contributed to British explanation of the difficulties in acceding to French request for forcing migrants from the bosom of kinsmen who were British subjects.

Among other considerations rendering cooperation inevitable between otherwise rival European colonial administrations was smuggling. The evolution of the intercolonial boundaries ultimately led to the development on both sides of intercolonial boundaries of two distinct market conditions. The variety of the goods available on both sides, and the differences in prices as one crossed from the one to the other, made transfrontier trade a real temptation. Smuggling of goods across the boundaries was a function of an endeavor on the part of local African business interests to avoid paying the chargeable customs duties and thereby to widen their profit margin.

The phenomenon, however, often gave European colonial administrations, as it has given succeeding African governments, much concern. First and foremost, the practice deprived the administrations of much-needed revenue. In a more serious sense, some aspects of smuggling were capable of causing the governments considerable embarrassment, such as the examples of arms transactions. Another was the flow of alcoholic beverages, coming into British territories mostly from the French sphere and for some time causing prohibitionist British authorities in Nigeria as much uneasiness as the arms and the ammunitions from their area were causing the French.[38]

But in spite of this situation which called for mutual vigilance, the persistence of smuggling throughout the colonial period and later points to some basic governmental inabilities. Part of the explanation stemmed directly from the deep-rooted jealousy between the different colonial administrations, which was to result in the contradiction implied in the official efforts at boundary control and the simultaneous establishment on either side of the intercolonial boundaries of rival rather than complementary market patterns and transportation systems designed deliberately to divert trade from the one side to the other.

The other side of the explanation related to the ingenuity of interested West Africans who did not fail to take full advantage of the weaknesses in the official control machinery. First, the local people had a more intimate knowledge of the terrain and of the countless footpaths, mostly sections of the precolonial trade routes, crossing the new frontier lines. This knowledge often proved a decisive advantage over that of the usually ill-equipped and frequently transferred customs officials operating in widely scattered stations. Besides, the frontier zone was generally neglected and lacking the access roads necessary for any effective patrol by the customs staff. Finally, the generally low salary structure in the customs services and the remoteness of their stations opened the officials to the temptation of bribes by the African business-men operating across the boundaries. For all the period of European rule and since, therefore, smuggling has continued both as a phenom-enon and as an element in interstate relations along the Nigeria-Benin and other West African borders.

To all the internal factors forcing interstate cooperation within West Africa one should add external considerations such as the German threat to the European international pact in 1914 to 1918 and again in 1939 to 1945; this threat further obliged the French and the British in West Africa, as elsewhere, to close ranks and work together to combat a common enemy. The conquest of German Togo by the Allied (British and French) forces during World War I was achieved largely by troops mobilized from Nigeria and Benin (Dahomey).[39]

Mechanisms for Binational Relations

To cope with common problems and interests, the colonial adminis-trations established formal channels of communication involving bilat-eral relations at the levels both of the territorial governments and their local administrations. The commonest of these took the form of visits and correspondences, on reciprocal basis, which were encouraged at the levels of both governors and local administrators. In 1923, for ex-ample, Governor Fourn of Dahomey paid an official visit to Sir Hugh Clifford, the governor of Nigeria, in Lagos.[40] Governor Clozel of the Ivory Coast had earlier visited the British governor of the Gold Coast. Of perhaps the greatest sensational effect was the visit of Sir Middleton of Gambia to Governor-General Carde in Dakar, at which occasion the visiting British governor conferred the title of the Knight of the British Empire on his French counterpart.[41] At each of these occasions, featuring lavish state dinners, luncheons, and other august receptions, official pronouncements were made with stress always on common problems and interests and the necessity for mutual cooperation.[42]

The governors of French and British colonies also exchanged infor-mation of mutual interest and benefits. The archives of former French West Africa are full of interesting dossiers containing budget drafts and speeches as well as copies of government gazettes of British West

African territories and vice versa. In pursuance of the same principle of cooperation and mutual understanding, the governor of every British colony regularly notified on reciprocal basis the French West African colonial authority of his appointment and assumption of office as well as of his departure on leave and the officer acting for him in his absence. The contacts between French and British West Africa through the communications between the governors were complemented through the functions and activities of the French and British consular mission in each other's colonial administrative headquarters in the subregion.

In the 1930s and 1940s, much encouragement was also given to regular meetings of local administrators, particularly those charged with frontier districts. Such meetings were held along the present-day Nigeria-Benin border between British District officers of Ilaro (later named Egbado) Division of Aboekuta Province, embracing the area of the various Western Yoruba communities located on the Nigerian side of the border, and the commandants of the adjacent circle, embracing the fractions of the same communities situated in French Dahomey. W. Duncas, British commissioner for the Western Province of the Gold Coast, gave a typical order for these local administrators' meetings as they applied to the Anglo-French relations along the border with the Ivory Coast when he wrote in 1933 to the district commissioner of Axim on the border:

> I regard it as most important that political officers in the frontier districts should be in close touch with, and maintain the most cordial relations with representatives of the neighboring power, in this case the French. You will, therefore, cooperate where possible with the French administrators across the frontier.
> In these days when the interests of Great Britain (and France) in West Africa are more or less bound together . . . and when many of the problems of administration and development are common to both governments, the need for cooperation . . . is sufficiently apparent.[43]

The obvious resemblance between the tone of this memorandum and the familiar rhetorics of today's West African leaders in respect to efforts at intergovernmental cooperation in the region is a good indication of the link between the colonial and postcolonial expressions of binational relations at the levels of the interacting sovereign governments.

Informal Relations Under Colonial Rule

It is clear from the foregoing account that most of the catalysts of bilateral relations between the interacting colonial governments arose directly or indirectly from the operation of policies within the ethnographical context of the border areas. The main issues were naturally connected with the local peoples' recognition of the border disequilibrium as manifested in observable differences between policies and policy effects produced by governments operating among the same people and the same localities. Armed revolts and especially protest migrations were typical borderland responses to obnoxious public policies. The fact of greater frequency of protest migratory movement to British sides of borders has been used as a valid measurement of the harsher nature of the French vis-à-vis British rule in West Africa.[44] The use of the Nigeria-Benin border in this respect fulfilled the typical role of borders as poles of attractions for nationals seeking political asylum.

Similarly, smuggling across the borders began as a function of the local traders' realization of the existence of two differential economies involving differences in the type and range of the commodities and their prices. On the Nigeria-Benin border, it began as a subtle form of resistance to an imposed border. Precolonial trade patterns, disrupted by a reorientation of the direction of trade and trade routes dictated by the new intercolonial state boundary, persisted. Smuggling grew in scope and scale in response to the growing realization of the profit factor as well as the differential size and rate of growth of the two colonial economies and improvements in transportation and communication systems.

It is not a defense of crime to point out that clandestine features of border transactions stem from the situation of borderlands as areas of states more prone to crimes than other subnational areas. This unfortunate circumstance results from what one might regard as an aspect of the paradoxes which Haddox has attributed to the border phenomenon viz, gross neglect and overcontrol by authorities of the interacting states.[45] Neglect manifests itself sharply on the Nigeria-Benin border characterized, like the other borders of Nigeria, as an unguarded boundary[46] by reason of few and widely distanced border control posts;

inadequately staffed, ill-equipped, and inappropriately motivated customs preventive service; a general remoteness from the main centers of intensive acculturation on both sides; and a virtual absence of all-season transportation systems whether across the border or between it and other areas of the state.[47]

Yet the border is overcontrolled. As Wheatman and Miller have explained with particular reference to the U.S.-Mexico situation,[48] laws and multiplicity of control mechanisms converge in unusual patterns. Not only are there the laws of both countries; international laws governing trade and passages between the two also exist. As in the U.S.-Mexico instance, law-enforcement on the Nigeria-Benin border is the responsibility of numerous, ill-coordinated agencies including the police, the customs preventive service, and the immigration authorities of both countries. From time to time the efforts of these agencies are augmented by detachments of the army and, with particular reference to Nigeria in the last seven years or so, the National Security Organization (NSO). The controls operate against the irresistible lure for high profit accruing from organized crimes such as smuggling. The situation poses a serious challenge to the adaptive capacity of borderlands business class. Survival and success under such conditions are locally regarded as positive achievements. Today, as in the colonial era, the situation on the Nigeria-Benin border is not different from what it has been on the U.S.-Mexico border where nationally recognized crime such as "smuggling ceases to be a crime, but identifies with the best part of the population and connects itself with the romance and legend of the border."[49]

The role of "crimes" in the life of border communities must not be exaggerated. In the colonial era, as today, not all border inhabitants were smugglers; one did not have to be a *fronterizo* in order to be a smuggler.[50] In the Nigeria-Benin borderlands, traders of dubious description constituted a numerical, though influential, minority.[51] The bulk of the population followed traditional peasant agriculture. Folk affairs dominated the thinking and activities of the vast majority. At the level of the community, the basic concern was for the survival of the collective consciousness and the maintenance of the interconnections both within and between the culture and subculture areas. In spite of the colonial partition, the Aja, the Yoruba, and the Borgu retained their cultural coherence and connections. This affinity should be a normal expectation for anyone aware of the fact that among these groups, as among most other African peoples, the past is a very vibrant part of the living present. Besides, the geographical contiguity of the subgroups, directly affected by the boundary arrangement with the rest of the larger culture area, in each case made the cause of cultural retention and group solidarity at the level an extremely easy process. From the riverain Gun to the savana-dwelling Borgawa, the entire area was devoid of obstacles to human mobility: the canoe between Badagry and Porto Novo and the numerous footpaths farther north have served the transport purpose of the communities. By the close of the colonial era, bicycles and the automobile had virtually rendered human porterage a thing of the past. Marriages, funerals, religious festivals, traditional musical concerts, and the *vodum* or traditional seminaries and centralized sanctuaries,[52] are among the several institutions pulling peoples within the distinct culture areas together. Of particular significance still was the impact of the traditional periodic markets.

Finally, traditional rulers and the politics of the chieftaincy institution exercised influence. This factor was (and is still) of particular importance to the Yoruba and Borgu areas. In Nikki where, as elsewhere in the French colonial sphere, traditional political institutions survived the colonial era in spite rather than because of French rule, succession to the kingship was throughout the colonial period a major issue in bilateral relations. Stewart has presented the picture very clearly:

> Whenever the throne of Nikki became vacant on the death of a king of Nikki, those border chiefs (in Nigeria) who were eligible for succession to the Nikki throne immediately pressed their claim. . . . The flurry of activity at such times had tremendous repercussions for the people. If, for example, a chief from Yashikera had been promoted to the Nikki kingship, his departure from Yashikera meant that several hundreds of his people would leave for Nikki with him. The (British) administration considered the constant migration back and forth across the frontier as disruptive to tax collection, and as hampering the efficient administration of these regions. The French colonial administrators of Northern Dahomey were faced with similar problems arising out of the creation of the international boundary as the British. Whenever the Chiefs of Yashikera (British subjects by definition) contested for the Nikki throne, the French were torn between putting on the throne the legitimate heir . . . but

realized if they did so they were undermining and weakening the argument for their right to rule their own region of Borgu.[53]

In spite of repeated failures, however, Yashikera-based candidates for the Nikki throne continued to strive for it throughout this period.

The sense of rejection of the border as a dividing line was far more explicit in the case of the Yoruba placed on the French side of the border. Adegbite Adewori, the alaketu (king) of Ketu, spoke for all of them in 1960 when, in response to a question posed by J.R.V. Prescott, a renowned geographer of the Nigeria-Benin border, he swiftly asserted that "We regard the boundary as separating the English and French not the Yoruba."[54] The alaketu has remained a very active influence in both Benin and Nigeria insofar as the traditional level of politics is concerned. From the colonial era to the present, his approval was customarily required before the headchiefs of other Ketu settlements in Nigeria and Benin could be validly installed.

The extent to which this influence has continued to be used for the achievement of overall solidarity of all Yoruba was evidenced by the January 1983 visit to Ketu of the ooni (king) of Ife, the Yoruba town in Oyo State of Nigeria, customarily believed to be the ancestral home of all Yoruba.[55] Earlier, in 1981, the traditional rulers of Nikki and Sabe were in attendance at the official installation of the okere (king) of Saki, a major Yoruba town in the northwest area of Oyo State of Nigeria. The presence of these Beninoi traditional rulers (one Borgu, the other Yoruba) was very significant for the analysis in this paper. Quite apart from the evidence provided for the survival of traditions of both intra-and interethnic relations of the Nigeria-Benin borderlands, the fact of the visit in the same week of the 1981 border incident, which provoked heated diplomatic exchanges between Nigeria and Benin, demonstrated the cleavage between the formal and the informal levels of border relations.

Binational Relations Since Independence

Since the attainment of sovereign status by Nigeria and Dahomey, details about binational relations have witnessed some changes. Armed revolts and protest migrations have, for example, ceased to be important issues, once French rule became steadily liberalized following the post-World War II reforms in French West Africa. The abolition of obnoxious policies such as forced labor, conscription, and excessive taxation helped to bring about some measure of parity with social conditions obtaining on the Nigerian, as on other British sides of common frontiers.

However, the more basic structural problems, giving rise to a state of socioeconomic disequilibrium, have remained. Thus, rather than bring about a harmonization, the era of independence was one of increased cross-border activities of a clandestine nature. The period since 1960 has witnessed a phenomenal growth in the scale of smuggling. Although a great deal of the transborder transactions are still concerned with imported manufactures, lorries and trailers have replaced head porterage and bicycles of the colonial era. A completely new dimension has been the bulk purchase and reexportation of Nigerian export agricultural produce. Here the scale of operation in respect to the Nigerian cocoa across the border with Benin has compared quite favorably with a similarly clandestine movement of Nigerian groundnut across the northern border into Niger.[56] Part of this innovation has been the direct involvement of the neighboring states of Benin and Niger, which officially provided the necessary infrastructure and, in turn, benefitted from the revenue accruing from fees and charges that were imposed.[57]

There has also been an intensification in the incidence of other border-induced crimes. Although there was already in the colonial period a regular flight of stolen bicycles from the Nigerian to the Dahomeyan side of the border, since the 1970s automobiles have been snatched by armed robbers. This has become a major menace in such Nigerian border states as Lagos, Ogun and Oyo, and an important factor in the clandestine transactions across the border with Benin.[58] The public safety aspect of the Nigeria-Benin border functions received heightened publicity recently when two top Nigerian politicians (both Yoruba from Oyo State), declared wanted by the Nigerian Police following the coup d'état of 31 December 1983, were reported to have escaped through the Yoruba section of the border.

Perhaps the most important single border issue in Nigeria, however, has been the question of "alien invasion" of Beninoi, Togolese, and Ghanaians who enter Nigeria "illegally," using channels and routes across the Nigeria-Benin border. Although economically induced migrations are known to have taken place back and forth across the border in the period before independence,[59] it was the Nigerian oil boom of the 1970s and the simultaneous collapse or near-collapse of the economies of neighboring states to the west, particularly Ghana, that brought about an unprecedented influx of Ghananian, Togolese, and other immigrants. The current Nigerian economic crisis, brought about by a combination of mismanagement by the ousted Nigerian government and global oil glut, led the government to decide on the massive repatriation of the aliens in January and February 1983. According to the constant newspaper reports about the steady back flow of the aliens,[60] the repatriation answer to the immigration question has proved as inadequate in Nigeria as it is known to have been in the U.S.-Mexico situation. In both situations, the ease with which undocumented immigrants merge with nationals of the same race and culture has helped in complicating the problem of detection by law enforcement agencies.

The importance of the informal network in all these activities remains obvious, yet government policy has persisted in the colonial tradition of ignoring the local social contexts. Not only were the colonial era contacts between neighboring border administrations discontinued by the postcolonial governments; Nigeria and Benin have maintained the tradition of dealing with each other as structurally differentiated entities in spite of evidence of strong ethnological links and the wide-ranging, complex, and interpenetrating activities across their common border. Rather than seek to cultivate consciously and adapt the informal networks and make them vital components of binational relations, the attitude of governments on both sides appears to be one of *laissez faire but watch.*

Nigeria and Benin operate as traditional good neighbors. Yet there have been occasional strains: (a) in 1960 when Dahomey was alarmed by suggestions of annexation of its Yoruba districts by some spokesmen of the Action Group of Nigeria, the political party in power in the predominantly Yoruba Western Region;[61] (b) during the Nigerian Civil War of 1967 to 1970 when Dahomey engaged itself in a series of anti-Nigeria activities; and, (c) most recently, in 1981 when the incursion of Beninoi *gendarmes* into the Nigerian side of the border in Sokoto State caused considerable controversy. But, by and large, the diplomatic relations have been cordial. Since the mid-1970s both countries have drawn more closely together. Indeed, the two have engaged in a series of joint ventures including the multi-million-dollar cement works at Onigbolo near Ketu in Benin and the sugar-processing plant near Sabe, also in Benin.

Although these enterprises, especially the cement works, now show signs of a metamorphosis into unilateral concerns,[62] the bilateral relations have remained quite healthy. A Nigeria-Benin Consultative Commission was set up following the border incident of 1981 and appears to be doing fine. Both countries are members of several common regional and subregional intergovernmental organizations, including the Organization of African Unity (OAU), founded in 1963, and the Economic Community of West African States (ECOWAS), inaugurated in 1975.

While the positive effects which excellent diplomatic relations at the level of sovereign states can have on border relations, including the informal networks, cannot be denied, it is important to stress that the survival of the informal relations is completely independent of the relations between the national authorities. Indeed, informal relations normally grow in spite of and not because of diplomacy. At any rate, the local-level initiatives on the Nigeria-Benin border have continued to flourish on the same bases as we have seen in respect to the colonial period.

Policy Advice

Given the enduring nature of the informal level of border relations and the conflict between it and the formal level as a major source of border irritations, the challenge for policy here is to establish the necessary harmony between the two. For the U.S. and Mexico, and more so for Africa, where border situations have been found to be more similar than dissimilar, attention must focus on the lessons of the Western European experience, where initiatives at the semiformal level, inspired by orientations of the informal networks, are now sanctioned by the sovereign nation-states through application of provisions of the European Outline Convention on Transfrontier Cooperation Between Territorial Communities or Authorities, passed by the Council of Europe in 1979, and fully ratified by 1982.[63]

In the U.S.-Mexico situation where the formal and semiformal levels of border relations are especially well developed, the main requirements will be the creation of appropriate mechanisms to enable the interrelating sovereignties to sanction the networks of relations already well developed between municipal governments across the boundary. However, in the proposed marriage between the formal and semiformal levels, the requisitive love, which comes from adequate attention to the strictly informal networks of relations between Chicano and Mexicano as "extended communities," must be induced and sustained.

In Africa, the challenge is to cultivate the informal level and incorporate it into the framework of local governments that can be trusted with the role of agents for binational relations. This need to adapt and formalize the informal linkages has been manifested on the Nigeria-Benin border as elsewhere in Africa where, as along Zambian borders with Mozambique and Malawi, Zambian government provisions of schools, medical care, and agronomic services for Chewa and Ngoni nationals have been infiltrated by kinsmen from the neighboring states' borderlands.[64] The task is to sensitize Africa's national governments and the appropriate intergovernmental organizations, particularly the Organization of African Unity, where the basic predilection is still for the outworn use of diplomacy and international law as main, if not sole instruments for the conduct of international relations.

What is called for is a genuine effort at international cooperation to lead ultimately to integrated planning and development at all levels, including those of the subnational areas along the borders. This is not an advocacy for the abolition of borders. It is an argument *for* future borders that will join and *against* those of today's nation-states, which seek to separate. It is a return to Boggs, whose advice has been "wholesome simplification of international boundary functions."[65]

Nigeria-Benin/U.S.-Mexico Border Comparisons

Items	Nigeria	Benin	Mexico	U.S.
Area in million sq. km.	.9	.1	2.0	10.0
Population (millions)	87.6	3.6	71.2	229.8
Population Location	Scattered		Scattered	
GNP Per Capita $U.S.	870	320	2,250	12,820
GNP (U.S. $ million)	76,212	1,152	160,200	2,946,036
Natural Endowment	Abundant	Moderate	Moderate	Super Abundant
Legal Tradition	British	Roman (French)	Roman (Spanish)	British
Government and Administrative Traditions	Federal and Decentralized	Centralist	Centralist States	Federal and Decentralized
Official Languages	English	French	Spanish	English
Dominant Western Christianity	Sectarian but predominantly Protestant	Pre-dominantly Catholic	Catholic	Sectarian but predominantly Protestant
Common Borders (miles)	About 500		2,013	
Greater Frequency of Border Crossings	Beninoi		Mexican	
Main Culture Areas Split	Borgu Yoruba Aja		Chicano/Mexicano	
Border City Development	Very Low		High	Moderate
Border Settlement Pattern	Interlocked		Interlocked	
Border Screening	Very High		Low	High
Migration	Moderate	High	High	Moderate
Customs	High	Low	Low	High
Smuggling	High	Very High	Moderate	Moderate
Smuggling (drugs)	Noticeable	Not so Noticeable	Very Low	High

Nigeria-Benin/U.S.-Mexico Border Comparisons

Items	Nigeria	Benin	Mexico	U.S.
Undocumented Immigration	Very High	Low	Very Low	Very High
Border Military Defense	Very Low	Low	Very Low	Very High
Border Research Cooperation by Government Agencies	NIL	NIL	Moderate	Very Low

Sources: A. I. Asiwaju, *Borderlands Research: A Comparative Perspective* (El Paso: Center for Inter-American and Border Studies, Border Perspectives Series, no. 6, November 1983), inspired by J.A. Price's table in E.R. Stoddard, et al., *Borderlands Sourcebook* (Norman: University of Oklahoma Press, 1983), 21. Population figures and GNP data are for 1981, as indicated in World Bank, *World Development Report, 1983* (New York: Oxford University Press, 1983), Table 1.

Notes

1. I am grateful to my home university, the University of Lagos, for making possible my participation in the 1984 El Paso conference and for supporting my comparative study of the Nigerian and U.S.-Mexico borderlands experiences. Thanks are also due to a host of American colleagues, particularly scholars at the Center for Inter-American and Border Studies at The University of Texas at El Paso and the Border Research Program at The University of Texas at Austin. I have also had a useful exchange of correspondence with Manuel L. Carlos of the University of California at Santa Barbara.

2. E. R. Stoddard, "Local and Regional Incongruities in Binational Diplomacy: Policy for the U.S.-Mexico Border," *Policy Perspectives Journal* 2:1 (1982): 112.

3. I. Duchacek, "Transborder Regional Linkages in the Era of Interdependence: A Conceptual Speculation" (Paper presented at the 1981 Conference of the Association for Borderlands Scholars), as summarized in *Frontera* 8:2 (1983): 11.

4. W. V. D'Antonio and W. H. Form, *Influentials in Two Border Cities* (Notre Dame: University of Notre Dame Press, 1965); J. W. Sloan and J. P. West, "Community Integration and Policies Among Elites in Two Border Cities: Los Dos Laredos," *Journal of InterAmerican Studies and World Affairs* 18 (1976): 451-74; J. W. Sloan and J. P. West, "The Role of Informal Policy Making in U.S.-Mexico Border Cities," *Social Science Quarterly* 58:2 (1977): 270-82.

5. L. Whiteford, "The Borderlands as an Extended Community," in F. Camara and R. Kemper, eds., *Migrations Across Frontiers: Mexico and the United States* (New York: State University of New York, 1979).

6. H. J. Briner, *Co-ordination of Regional Planning across National Frontiers, Regio-Test Case: Switzerland/France/Germany* (Basel: Regio Basiliensis, 1981); N. Hansen, "European Transboundary Cooperation and Its Relevance to the United States-Mexico Border," *APA Journal* (Summer 1982): 336-43; A. I. Asiwaju, "Borderlands as Regions: Lessons of the European Transboundary Planning Experience for International Economic Integration Efforts in Africa" (Paper presented at the Invitational International Seminar on the Economic Community of West African States [ECOWAS] and the Lagos Plan of Action, the Nigerian Institute of Social and Economic Research [NISER], University of Ibadan, Ibadan, 12-16 December 1983).

7. E. R. Stoddard is distinguished in his perceptive studies of the localized impact of the U.S.-Mexico border. However, the fact of a virtual absence of systematic research on the sociology/anthropology of the border is indicated in M. L. Carlos' "Toward an Anthropology of Border Chicano/Mexicano Society: A Pragmatic Approach," *The New Scholar* (1983). An excellent research for a radio program, based on the concern for the humanistic dimension, is contained in the following report: V. S. Wheatman and T. E. Miller, *Final Report "On The Border,"* The National Endowment for the Humanities, The University of Texas at El Paso, Public Radio Station KTEP. (Copy in the Border Studies Collection, Special Collections Department, UTEP Library.)

8. S. W.Boggs, *International Boundaries: A Study of Boundary Functions and Problems* (New York: Columbia University Press, 1940), 96-133; V. F. Malchus, *The Cooperation of European Frontier Regions* (Strasbourg, Council of Europe, 1975), 2-3; S. A. Beagle, H. F. Goldsmith, and C. Loomis, "Demographic Characteristics of the U.S.-Mexico Border," *Rural Sociology* 25 (1960): 107-62.

184 • ACROSS BOUNDARIES

9. J. C. Anene, *The International Boundaries of Nigeria, 1885-1960: The Framework of an Emergent African Nation* (London: Longman, 1970), 141-232; J. C. McEwen, *The International Boundaries of East Africa* (New York: Oxford Univerity Press, 1971); S. Touval, *Boundary Politics of Independent Africa* (Cambridge: Harvard University Press, 1972); C. Widstrand, ed., *African Boundary Problems* (Uppsalla, 1969); I. Brownlie, *African Boundaries: A Legal and Diplomatic Encyclopedia* (London: C. Hurst, 1979).

10. See the following works by A. I. Asiwaju: *Western Yorubaland Under European Rule, 1889-1945: A Comparative Analysis of French and British Colonialism* (London: Longman, 1976); "The Aja-Speaking Peoples in Nigeria: A Note on their Origin, Settlement and Cultural Adaptation up to 1945," *Africa* (International African Institute in London) 49:1 (1979): 15-28; and "Border Populations as a Neglected Dimension in Studies of African Boundary Problems: The Nigeria-Benin Frontier Case" (Paper presented to the Invitational Seminar on Nigeria's International Boundaries, Nigerian Institute of International Affairs, Lagos, 5-7 April 1982).

11. A. I. Asiwaju, *Borderlands Research: A Comparative Perspective* (El Paso: Center for Inter-American and Border Studies, Border Perspectives Series, no. 6, November 1983).

12. C. W. Newberry, *The Western Slave Coast and its Rulers: European Trade and Administration among the Yoruba and Aja-Speaking Peoples of South-Western Nigeria, Southern Dahomey and Togo* (New York: Oxford University Press, 1961); Anene, *The International Boundaries of Nigeria*; J. R. V. Prescott, *The Evolution of Nigeria's International and Regional Boundaries, 1861-1971* (Vancouver: Tantalus Research, 1971); Asiwaju, *Western Yorubaland*; A. Mondjannagni, "Quelques Aspects Historiques, Economiques et Politiques de la Frontière Dahomey-Nigeria," *Etudes Dahoméenes* (Nouvelle Série, 1963): 17-58; L. R. Mills, "An Analysis of the Geographical Effects of the Dahomey-Nigeria Boundary" (Ph.D. diss., University of Durham, 1970).

13. Anene, *The International Boundaries of Nigeria*, chapters 5 and 6.

14. Asiwaju, *Borderlands Research*.

15. V. Thompson, "Dahomey," in G. M. Carter, ed., *Five African States: Responses to Diversity* (Ithica: Cornell University Press, 1964), 250-52.

16. See the following works by J. O. Igue, "Un Aspect des Echanges Entre le Dahomey et Le Nigeria: Le Commerce du Cacao," *BIFAN*, tome 38 série B (1976): 636-69; "Evolution du Commerce Clandestin Entre Le Dahomey et Le Nigeria Depuis du Guerve du Biafra," *Canadian Journal of African Studies* 10:2 (1976): 235-57.

17. W. H. Scotter, "International Rivalry in the Bights of Benin and Biafra" (Ph.D. diss., University of London, 1933); A. B. Aderibigbe, "The Expansion of the Lagos Protectorate Frontier, 1863-1900" (Ph.D. diss., University of London, 1959); Newbury, *The Western Slave Coast and its Rulers*; J. D. Hargreaver, *Prelude to the Partition of West Africa* (New York: Macmillan, 1963).

18. Anene, *The International Boundaries of Nigeria*, 141-232.

19. Mondjannagni, "Quelques Aspects Historiques," 25.

20. Brownlie, *African Boundaries*, 167-71.

21. K. T. Opuku, "Our Civil Law Neighbours," *University of Ghana Law Journal* 4:1 (1967): 40-53.

22. J. House, *Frontier on the Rio Grande* (New York: Oxford University Press, 1982), chapter 8.

23. A. I. Asiwaju, "The Aja-Speaking Peoples in Nigeria: A Note on Their Origin,

Settlement, and Cultural Adaptation to 1945," *Africa* (International Institute of London) 49:1 (1979): 15-28.

24. D. Forde, *The Yoruba-Speaking Peoples of South-Western Nigeria* (London: IAI, 1951); J. S. Eades, *The Yoruba Today* (London: Cambridge University Press, 1980).

25. J. Bertho, "La Parente des Yoruba aux Peuplades du Dahomey et du Togo," *Africa* 19 (April 1949): 121-32; P. Mercier, "Notice sur le Peuplement Yoruba du Dahomey-Togo," *Etudes* Dahoméenes (E.D.) 4 (1950): 29-40; E. G. Parrinder, "The Yoruba-Speaking Peoples in Dahomey," *Africa* 7:2 (1947): 122-28; E. G. Parrinder, "Some Western Yoruba Towns," *Odu* 2 (1955): 4-10.

26. M. Crowder, *Revolt in Bussa* (London: Faber and Faber, 1974); M. Stewart, "The Borgu People of Nigeria and Benin: The Disruptive Effect of Partition on Traditional Political and Economic Relations," to be published in the *Journal of the Historical Society of Nigeria*.

27. B. A. Agiri, "Yoruba Oral Tradition with Special Reference to the Early History of the Oyo Kingdom," in W. Abimbola, ed., *Yoruba Oral Tradition in Music, Arts and Drama* (Institute of African Studies, University of Ife, 1974); R. C. Law, "The Northern Factor in Yoruba History, in Akinjogbin and Ekemode, *Proceedings* of the Conference on Yoruba Civilization; Asiwaju, "The Aja-Speaking Peoples in Nigeria."

28. Akinjogbin, "Toward a Historical Geography of Yoruba Civilization."

29. This point has contributed immensely to the difficulty in distinguishing between Nigerian and non-Nigerian nations when both relate to community with an indigenous base in Nigeria. Alex Quaison-Sackey has dramatized the same point with a little bit of journalistic exaggeration when he states that "All West Africans are citizens of West Africa. They were even before the promulgation of ECOWAS. From Mauritania to Nigeria, the people crisscross linguistically and culturally. Thus Nigerians have relations in Benin who have relations in Togo, who have relations in Ghana, who have relatives in Liberia, who have relatives in Sierra Leone and so on." *West Africa*, 21 February 1983: 47. This problem has provided the basis of a celebrated court case in Nigeria. See Suite BOM/13M/80: "Shugaba Abdulrahman Daram (Applicant) v. The Federal Minister of Internal Affairs and Others (Respondents), High Court of Borno Judicial Division, Maiduguri, 1980. They contested the politically motivated expulsion of the plaintiff from Nigeria on the allegation that he was not a Nigerian, based on the fact of family history links which the plaintiff as a Kanuri (a Nigerian ethnic group fragmented by the Nigeria-Chad-Cameroon border) has with Chad. The case naturally enlisted much interest in Nigeria particularly among Yoruba peoples of Ogun, Lagos, Oyo, Ondo, and Kwara states, who like the Kanuri, are split by a border (eleven out of Nigeria's nineteen states have border locational character).

30. The bulk of the discussion is derived from A. I. Asiwaju, "The Socio-Economic Integration of the West African Sub-Region in the Historical Context: Focus on the Colonial Period," Bulletin de *L'I.F.A.N.* *(BIFAN)*, Dakar Senegal, série B tome 40, no. 1 (1978).

31. Asiwaju, *Western Yorubaland*, 196; Mondjannagni, "Quelques Aspects Historiques," 27.

32. A. I. Asiwaju, ed., *Partitioned Africans: Ethnic Relations Across Africa's International Boundaries, 1884-1984* (London: C. Hurst and Company, 1984), Appendix.

33. Asiwaju, *Western Yorubaland*, 146.

34. A. I. Asiwaju, "Anti-French Resistance Movement in Ohori-Ije (Dahomey), 1895-1960," *Journal of the Historical Society of Nigeria* 7:2 (1974).

35. Crowder, *Revolt in Bussa*.

36. A. I. Asiwaju, "Migrations as an Expression of Revolt: The Example of French West Africa Up to 1945," *Tarikh* (Journal Published for Schools and Colleges by Longman for the Historical Society of Nigeria) 5:3 (1977).

37. A. I. Asiwaju, "Migrations as Revolt: The Example of the Ivory Coast and the Upper Volta Before 1945," *Journal of African History* 17:4 (1976): 577-94.

38. A. I. Asiwaju, *Western Yorubaland*, 196-99.

39. J. Osuntokun, *Nigeria in the First World War* (London: Longman, 1979).

40. Archives Nationales du Senegal (ANS), Dakar, série BG 53(32): Visit of the Lieutenant Governor of Dahomey to Lagos, 12-13 January 1923 (Lagos Government Printer, 1923).

41. Archives Nationales du Senegal: série 5F.7: Visit of Lieutenant Governor Clozel of the Ivory Coast to Mr. Nathan, British Governor of the Gold Coast (present-day Ghana).

42. Archives Nationales du Senegal: série 1F 5 (14): Governor H. R. Palmer of the Gambia to Governor-General Brevie in Dakar, Barthust (now Banjul), 31 October 1930; Agence Consulaire de France at Barthust, to Governor-General in Dakar, 10 November 1931.

43. Archives Nationales de la Côte d'Ivoire, Abidjan: Standing Order No. 3/1933 at Sekondi by British Commissioner for the Western Province of the Gold Coast, to the District Commissioner of Axion.

44. Asiwaju, *Western Yorubaland;* Asiwaju, "Migrations as Revolt"; Asiwaju, "Migrations as an Expression of Revolt."

45. J. Haddox, "The Border: A Place to Live, A Place to Learn" (Faculty Research Lecture, The University of Texas at El Paso, 1982).

46. J. Ekpeyong, "Nigeria's Unguarded Boundaries," *Forum, A Journal of the Nigerian Institute of International Affairs* (1982).

47. Mills, "An Analysis of the Geographical Effects of the Dahomey-Nigeria Boundary," 44-45.

48. Wheatman and Miller, *Final Report "On the Border."*

49. William Emory's Journals, as quoted in Wheatman and Miller.

50. Igue has shown, e.g., that the catchment area for the cocoa smuggled across the Nigeria-Benin border came from as far away as Ibadan, Ife, and Ondo in the heart of the Cocoa Belt in Western Nigeria and that Beninoi entrepreneurs were drawn from as far as Cotonou, Whydah and Agbome (Abomey). Igue, "Un Aspect des Exchanges."

51. Mondjannagni, "Quelques Aspects Historiques," 51-53; Igue, "Un Aspect des Exchanges," 654-63.

52. Asiwaju, "The Aja-Speaking Peoples in Nigeria," 21-24.

53. M. Stewart, "The Borgu People of Nigeria and Benin."

54. Prescott, *The Evolution of Nigeria's International Boundaries*, 103.

55. A report on this visit was filed by the present writer in the issue of *West Africa* (a widely circulated London-based weekly magazine), 28 February 1983. On this visit to Ketu, the ooni of Ife was accompanied by such other Nigerian Yoruba Oba as Orangun of Ila, Timi of Ede, Eleruwa of Eruwa and Akran of Badagry. With the alaketu on this occasion were all traditional crowned heads of Ketu land east and west of the border as well as the onisabe of Sabe. A highlight of the visit was the conferring of minor chieftaincy titles on about 100 prominent Nigerians and Beninois of Ketu descent by the alaketu. The visit itself reciprocated two previous ones which the alaketu had paid to the ooni, one at Ife in 1980 and the other in 1982, to attend the Conference of Yoruba Oba meeting on Ibadan at the instance of the ooni of Ife.

56. Igue, "Un Aspect des Exchanges"; Igue, "Evolution du Commerce Clandestin," J. D. Collins, "Clandestine Movement of Groundnut across the Niger-Nigeria Boundary," *Canadian Journal of African Studies* 10:2 (1976): 259-76.

57. Igue, "Un Aspect des Exchanges," 637-43.

58. Asiwaju, *Western Yorubaland*, 197. See C. Oluduro, "Auto-Theft in Nigeria," a researched survey serialized in the *National Concord* (one of Nigeria's most respected private newspapers), issues of 10-13 October 1983. For a comparison with the U.S.-Mexico border, see "Auto Theft: Where did My Car Go?" in the maiden issue of *Borderlands*, a magazine of the Borderlands Project of the El Paso Community Collge, inserted into *The El Paso Times* 14 August 1983, and Sloan and West, "The Role of Informal Policy-Making," 277-78.

59. Mondjannagni, "Quelques Aspects Historiques," 31.

60. A typical report was "Ghananians Are Coming Back: Despite Crime and Poverty, Lagos Draws Many Sent Home in January," in *Sunday Concord*, 9 October 1983. Compare with the more scholarly study of Jorge Bustamante, curiously titled "The Mexicans Are Coming: From Ideology to Labor Relations," CEFNOMEX, Tijuana, Mexico (n.d.).

61. Thompson, "Dahomey," 252.

62. This information was gathered during discussions at the NISER Seminar in December 1983. See Asiwaju, "Borderlands as Regions."

63. Council of Europe, *European Outline Convention of Transfrontier Cooperation between Territorial Communities or Authorities*, no. 106 of the European Treaty series (Strasbourg, 1982).

64. Asiwaju, "Neglected Populations"; S. H. Phiri, "Some Aspects of Spatial Interaction and Reaction to Governmental Policies in the Border Area: A Study in the Historical and Political Geography of Rural Development in the Zambia-Malawi and Zambia-Mozambique Frontier Zones, 1870-1979" (Ph.D. diss., Liverpool, 1980).

65. Boggs, *International Boundaries*, 133.

Part VI

Communist Borders

Border Problem Solving in the Communist World: A Case Study of Some European Boundaries

Z. A. Kruszewski

Border problem solving in the Communist world is a subject not adequately covered in the literature of political science or, more specifically, in that of international relations. One can find many references to the particular border problem-solving strategies of individual Communist countries in either highly specialized journals published in each particular language[1] or in scattered references to the subject published in the daily press of each country.[2] The subject, however, has not found its author in a comparative cross-national presentation and still awaits an appropriate forum and interest.

This preliminary discussion addresses some basic principles affecting the Communist borders, owing to the peculiarities of the political system, and only attempts to list some similarities and differences both between various Communist borders and especially between the Communist and non-Communist ones.

Western scholars, even political scientists, are often unaware of the vast differences to be found in this subject matter within the Communist world, for while most are familiar with the Berlin Wall border, they may assume similar conditions exist on other Communist borders. In reality the various borders of the sixteen Communist states of the world constitute a wide variety of sociopolitical circumstances which create an array of specific border problems, especially when those borders face the non-Communist countries.

This discussion will focus on (a) the European-Communist borders in one region only, and (b) a few aspects of border problem solving and the type of problems encountered. The region that will be the subject of our attention is the so-called "northern tier" countries composed of the German Democratic Republic, Poland, and Czechoslovakia, with some emphasis on the East-West border running through Germany and that dividing Eastern Europe from the USSR (or the European Soviet bloc countries and the USSR). Only a few, very select problems will be addressed, since the subject in its entirety requires in-depth study.

Early Marxist theory tended to overlook the role of borders, largely because of the emphasis on world revolution: the establishment of the self-governing grassroot units of production (large and small) would replace the bourgeois nation-states, which would "wither away." The world revolution did not occur, however, and the Great October Revolution of 1917 initially resulted in establishing two Communist states (theoretical irony) of the USSR and Mongolia.

Therefore, the problem of borders and border problem solving had arisen but was cast in a uniquely hostile framework — a twilight zone of neither war nor peace — because of the construction, after 1921, of the "cordon sanitaire" along the western border of the USSR by France and her Eastern European allies — Finland, the Baltic States (Estonia, Latvia), Poland, and Romania.[3] Furthermore, the USSR was supposed to be henceforth the only permanent Communist nation-state of global nature later to absorb all other nations through the success of the world revolution to come.

The post-World War II expansion of Communism brought forth another ironic twist to Marxist theory, either original or its Stalinist version: the birth of an additional fourteen Communist nation-states, which changed the assumption mentioned above. These states, which began to function in a much more peaceful world after 1945 (especially in Europe), introduced the question of relations between each neighboring state, as well as that of the relations along the Communist and non-Communist borders, now much more open to traffic of goods and people than in the interwar years between World War I and World War II.

Several developments affected these borders, and the relations along them varied. The advent of cold war, division of Germany, secession of Yugoslavia and Albania from the Soviet bloc, and many destabilizing internal political tremors within the bloc (East Germany, Poland, Hungary, Czechoslovakia, and Poland several times later) were some of the elements affecting, among other things, the borders and the relations along them.

Furthermore, polycentric developments within Eastern Europe, initially triggered by Yugoslavia and later copied in milder versions by the Hungarians, Poles, and even the East Germans, loosened not only the previously rigid relations on their borders but even those facing the West. These borders largely ceased to function as absolutely sealed dividing lines. It is interesting to note that although the east-west border across Germany had changed considerably since the German treaties of 1972 and had opened up,[4] albeit in a one-sided way, the only border that remained relatively unchanged was the Soviet border with Eastern Europe, facing Poland, Czechoslovakia, Hungary, and Romania.

Before considering some of the border problem solving in this part of Communist Europe, one aspect of border relations resulting from World War II — the vast exchanges and/or expulsions of populations — is very salient. Whereas all borders in the world are admittedly unjust essentially and usually result from power relations — however recent or removed in time — they have always brought about some cross-migration of population unhappy with the change of sovereignty or forced relocation under the new authority.

Never before in modern history, however, have such population movements been effected as during and after World War II in Poland, Germany, and the USSR. The boundaries between those states were not set according to the previous practice (vainly attempted by Wilsonian diplomacy) to delineate the extent of ethnic and/or linguistic areas. Instead, the nation-states were to conform to the new boundaries.

Thus, we have had the first such episode in the modern history of Europe of bodily moving Poland (the sixth largest European state) some 150 miles west on the map of Europe, resulting in the migration of one-third of the Polish nation.[5] Similarly, some 16 percent of the Germans were either transferred or expelled from East Europe, and 10 percent of all Finns had to move within the new borders of Finland. Czechs and Slovaks also had to leave the territory they lost to the USSR.

Thus, almost all the Soviet western borders (except those facing Hungary and Romania) were lines now clearly separating different peoples and severing contacts previously established. The same was true along the new Polish-German border on the Oder-Neisse rivers, clearly dividing newly settled Poles from the Germans.

A host of problems arose with these massive relocations. All borders, established in the new arbitrary but clear-cut way, prevented not only normal relations along them for many years but prohibited any contact at all. Mundane problems like lost cattle, pollution, or emergencies along the borders had to be solved at the highest level in respective capital cities, despite treaties regulating such situations — treaties entered into at the time the borders were established.[6]

These problems especially affected the East-West German border, since no formal recognition of the division was granted by either German state until 1972; hence, no border problem-solving policies existed, nor were informal ones allowed to operate in a regular manner. Mechanisms subsequently put into operation are now fairly effective, easing life along the border and restoring personal contact, although movement from west to east, except among retirees, is far more prevalent than its reverse.[7]

Border problem solving in the Communist world is not all of a piece, however. While detente and the West German "Ostpolitik" have established some contacts across the intra-German border, polycentric developments within Soviet-bloc countries, particularly Yugoslavia, Poland, and Hungary, loosened relations. Hence, presently a correlation exists between the ideological and political rigidity of a given regime and the porosity of its borders.[8]

Historically, these variations in across-border contact developed only over time. Immediately following World War II, there was a great deal of fluidity across borders, but once Communist "law and order" was established, all the borders of eight Eastern European countries and the USSR were frozen. Henceforth contact across those lines was impossible for unauthorized persons, highly restricted even for officials, and initially open to only a handful of tourists at a few selected crossing points. Treaties regulating contacts along and across the borders remained largely moot, since the three-to-five kilometer-wide security zones along the boundaries were prohibited to all but border police and local population.[9]

Furthermore, an immediately adjacent zone, one-half to one kilometer from the border, was cleared of all buildings (save police and military ones) and people. Special permits were given for the cultivation of fields under guard or for the collection of wood. The borders of all the Communist countries were cleared by cutting border strips through forests and/or removing buildings. The zone was fortified —

with watch towers, housing machine guns, and artillery bunkers — and closed with barbed wire (if no walls existed).

These conditions prevailed on all Communist borders at least until Stalin's death and the beginning of the "thaw" in the mid-1950s, if not later until the mid-1960s "detente." They still exist largely along the East-West intra-German border, along the USSR-East European borders, and in other selected places.

Many border restrictions have either relaxed or been swept away subsequently by detente and polycentric developments in various Communist countries and the enormous growth of tourism between Communist countries and non-Communist countries.[10] Zones along the borders, in especially attractive areas, were opened after Communist authorities discovered the value of tourism to the Gross National Product. Areas such as the Mazurian Lake district in Northern Poland, the Bieszczady Mountains in Southeast Poland (empty since the expulsion of all Ukrainians), the Sudetic Mountains, and the Bohemian Forest (cleared of Germans) are some examples. Special tourist zones thus were opened along the borders, accessible to the citizens of bordering countries. This relaxation of rules pertaining to tourists enabled repopulation of these areas by either new settlers or some returning inhabitants. Contacts up to the border and even across the border were allowed also. The previously signed treaties pertaining to the border problem solving finally could become operational.

As soon as relaxation began, the polycentric nature of Communist border policy became evident. Contacts along and across the borders were now regulated and encouraged on a large scale between East Germany, Poland, and Czechoslovakia (clearly nothing happened along the Polish-USSR border), and between Austria, Hungary, and Yugoslavia, with the latter establishing fairly open borders once it allowed large-scale emigration of citizens to Western Europe in the 1960s. Yugoslav borders are different from other Eastern European ones, as they now are almost completely open to all traffic and contacts.

The most ambitious border-opening experiment in Eastern Europe existed between Poland and East Germany, from January 1971 until the beginning of the Solidarity Movement in Poland in August 1980. The Polish-East German border along the Oder-Neisse rivers was opened to both Poles and Germans, permitting, for the first time in the Communist world, entry into each other's country without visas. In the first year 6.7 million Germans (almost half of the East German population) visited Poland, and 9.5 million Poles (almost one-third of all

Poles) visited East Germany.[11] Several new border crossings were established, and contacts across the border during the 1970s resembled the U.S.-Mexico border, with Poles buying ready-made clothes and appliances in Germany and Germans buying gasoline and eating meals in Poland. The Solidarity Movement frightened East Germans into terminating these contacts to a great degree.[12] Even before 1980, the mass influx of Polish buyers caused severe shortages of appliances and resulted in several currency regulations that limited the volume of allowed purchases.[13]

Apart from the above examples of the open intra-Communist border, the contacts across most of the Communist borders can be largely classified into three categories: political, economic, and cultural. The political contacts encompass not only officials crossing the border but also "friendship delegations" to celebrate a particular event or situation. Crossings are then allowed even in places not normally open, but only on a one-time basis. Officials also cross to solve some particular problem affecting the border between the countries. The genesis of these political contacts was the state-to-state and local agreements which were entered into once relations became more relaxed, especially between East Germany, Poland, and Czechoslovakia in the region considered in this paper.

The easing of previous restrictions and the "discovery" of the economic benefits of tourism transformed some of these Communist borders to a great degree. Many more crossings were opened: East Germany-Poland (287 miles), from three in 1971 to eleven today; Poland-Czechoslovakia (820 miles), from three to eight; and East Germany-West Germany, from ten to fifteen. The visa-issuing procedure was simplified for Western tourists and abolished for tourists from other People's Democracies. Border crossing for employment was yet another development, both along the East German-Polish border, and, to a lesser extent, along the Polish-Czechoslovakian one. In the former case, for example, some ten thousand Poles, mostly women, much as in the U.S.-Mexican case, now cross the border daily to work in the textile and electronic industries in several German border cities. Smaller numbers cross into Czechoslovakia to work in Silesian industry. These contacts opened in the early 1970s and, incidentally, brought thousands of Polish-German marriages, a hopeful symbol after the tragic memory of World War II.

The negative aspects of economic development, especially along the German/Polish and Czechoslovakian/Polish borders, were widespread probems of pollution of water and rivers. The extent of the damage done alarmed the press in those countries and forced them to sign and implement, sometimes reluctantly in the face of economic benefits, special antipollution laws.

The cultural contacts are especially fostered across various Communist countries (in Communist usage, "cultural contacts" include some aspects of organized youth-oriented tourism/hiking, vacation, and sports). They are usually mass-organized and large scale. They involve folklore groups with local exchanges across the border or national ones visiting the neighboring country.

These contacts, in all three forms — political, economic, and cultural — enable the citizens of all Communist countries to breach the otherwise impenetrable borders of their own countries. For instance, the divided West and East German families traditionally meet for vacations in Poland and Czechoslovakia or on the beaches of Bulgaria, often fleeing to the West after obtaining documents there. East Germans, Lithuanians, and other Soviet citizens have either traveled in the past to Poland to experience "the West" or read the Polish press (for years *Zycie Warszawy*, a leading Warsaw paper, was published also in German for the East German tourists until stopped by demand of the East German government, since it obviously carried information from the West never published in the East German press).

In spite of the slight opening of the East European borders, some border problem peculiarities remain. All Communist countries, except for Poland, Hungary, and Yugoslavia, generally forbid emigration of their citizens. Hence citizens frequently attempt illegal crossing of borders (especially, the intra-German border but also Czechoslovakia and for a while after the imposition of martial law, Polish borders facing west). The ingenuity of those smuggling themselves out of those countries and of large-scale smuggling gangs is unlimited; their activities are persistent even during the strictest political periods and in spite of long jail sentences (or death in the case of East German escapees).

Elaborate efforts of various Communist governments have been made to stop the flourishing smuggling of various goods and scarce articles across their national boundaries.[14] In spite of the strict inspections on both sides of the border, it not only goes on unabated but actually increases. The press of all those countires is full of news about smashed smuggling rings worth millions of dollars. Corruption affecting all levels of Communist governments is largely responsible for the impossibility of eradicating large-scale smuggling (e.g., cameras and machines to Poland, dollars and other hard currencies from Poland to the West, icons and antiques from the USSR to Poland, leather from Bulgaria to Hungary, Czechoslovakia, and Poland).

Although there is a general tendency toward the stability of international relations in post-Helsinki Europe, and practically all the Communist borders are more open than in the past, the Soviet border is rigidly controlled and practically no contact is allowed except for some politically organized celebrations and visits. The border areas on the Soviet side are furthermore inaccessible to visitors from the outside (not only *Kaliningrad oblast* but also western parts of Belorussia and the Ukraine). That part of the Soviet border with Poland, some 780 miles long, has had only two road and three railway crossing points since 1945 and is closely guarded (like the intra-German border) by the Soviet watch towers. The border with Romania, 864 miles long, has only three road and three railway crossings. It is indeed a miracle of endurance and wit to be able to engage in large-scale smuggling under such circumstances.

These preceding preliminary comments raise more questions than they settle and only testify to the need for the systematic study of Communist borders in the comparative perspective of cross-national studies. This attempt only suggests the wealth of subject matter that has not been explored and which offers to prove fruitful in research addressing the larger theoretical issues of international relations and border studies.

Notes

1. For example, Karin Schmid *Dad Grundgesetz der DDR Teil I: Zusammenhange mit dem Grenzgestz der Ud SSR?* Berichte des Bundesinstitut fur ostwissenschaftliche und internationale studien, Koln 37/1983.

2. Recent articles in a leading Warsaw weekly *Polityka* reported on transborder traffic of the Rumanian tourists who, while visiting Yugoslavia, take up jobs for which there is no demand among the unemployed Yugoslavs because of low wages. *Polityka* 5 (1396) 4 April 1984.

3. Isaiah Bowman, *The New World Problem in Political Geography* (Chicago: World Book Co., 1924); Derwent Whittlesey, *The Earth and the State* (New York: Holt and Co., 1939).

4. Federal Republic of Germany, Press and Information Office, *Documentation Relating to the Federal Government's Policy of Detente* (Bonn, 1978).

5. Z. Anthony Kruszewski, *The Oder-Neisse Boundary and Poland's Modernization* (New York: Praeger, 1972).

6. For example, Andrzej Lesniewski, ed., *Western Frontier of Poland. Documents, Statements, Opinions* (Warsaw: Polish Institute of International Affairs/Western Press Agency, 1965), 90-92.

7. Federal Republic of Germany, Press and Information Office, *A Mandate for Democracy* (Bonn, 1980), 161-63.

8. *Polityka.*

9. In Poland those security zones were not dismantled until 1957, following the 1956 "Polish October." In other countries, i.e., Albania, USSR and German Democratic Republic, they have been functioning continuously since World War II (in the USSR, since the 1920s).

10. *Rocznik Statystyczny 1983*, year XLIII (Warsaw: GUS, 1983), table 140 (942).

11. *Rocznik Statystyczny 1974*, year XXXIV (Warsaw: GUS, 1974), table 10 (788).

12. It is being slowly reactivated, albeit on a very small scale. See *Krajowa Agencja Informacyina*, 50/1204/XXIV (12-18 December 1983), 3.

13. The movement was widely reported in the Polish and East German press at the time.

14. "Guestworkers," in *Polityka* 14 (1405) year XXVIII, 16. The article pokes fun at widespread, mass smuggling of scarce articles by the Poles visiting East Germany on a daily basis.

Concluding Observations

Oscar J. Martínez

With the growth of interdependence across international boundaries around the world, many nations have found it necessary to establish frameworks for transborder cooperation and planning. Economic, social, cultural, and demographic problems that were absent along border frontiers in earlier periods because of sparseness of population and remoteness from core areas must now be addressed to maintain good international relations. As the essays in this volume point out, however, solving border problems is a complex and often frustrating process. Relatively few countries that share borders with one another are favored with the conditions that make possible effective problem solving in binational zones.

Fruitful border cooperation is most likely to occur in cases where neighbors are no longer concerned with territorial disputes or other destabilizing situations, and where the economic and material lifestyles of their populations are not appreciably dissimilar. Clearly Western Europe has achieved these favorable conditions, with the consequence that many nations in the region have made impressive progress in achieving productive transborder cooperation.

By contrast, the United States and Mexico have been unable to tackle their border problems effectively. Among the factors that have worked against the development of significant border cooperation between Americans and Mexicans are the great disparity in the wealth of the two nations and the population pressures within Mexico. Economic hardship has driven millions of people from the interior of Mexico to the northern border and into the United States, causing Americans to feel threatened with perceived "invasions" of foreigners. As long as the economic and demographic gap prevails along the U.S.-Mexico border, it will be very difficult to reach agreements on meaningful joint border programs. The U.S. government in particular is unlikely to take any steps toward undertaking projects that might stimulate even greater cross-border migration.

As Mexico improves its economy and lowers its rate of population growth, as it has been doing of late, perhaps a more favorable climate will develop that will lead to the creation of the kind of organizations that exist in Western Europe to handle border problems. It is likely, however, that the changes required of Mexico will not occur in the next generation, and therefore truly significant border cooperation will be delayed for many years to come. It is apparent that the two nations will continue to pursue very specific agreements to resolve problems with minimum social or economic implications, such as maintaining a stable river boundary and cleaning up contaminated water along the border.

Regardless of the border issue to be addressed and the nation-states involved, one important point needing emphasis is that border people themselves must play a more active role in the process. That is one of the important lessons arising out of the border experience of Mexico and the United States, and the point is strongly reinforced by the record in frontier regions in Western Europe. Solutions have a greater chance of achieving permanence if borderlanders are allowed by their respective central governments to take part in a meaningful way in any transboundary negotiations that take place. Federal officials must realize that borderlanders are more knowledgeable than they about the nature of border problems and are in a better position to identify workable solutions to complex local problems. The most intense interest in regional issues is found in the region itself, and successful implementation of policy depends to a significant degree on the level of commitment to it by local populations.

It is unfortunate but true that estranged relations often exist between borderlanders and federal authorities. Adverserial attitudes seem to develop from the presence of the nation state in border areas in the form of coercive agencies that tightly control the movement of people and goods across the boundary. For a border region to flourish, people who live in the area must not be impeded from crossing back and forth from one country to the other, and a reasonable amount of trade must be permitted to take place. Restrictions are seen by borderlanders as unfriendly acts against them, and serve to reinforce the view that central governments lack understanding and sensitivity to their unique situation as residents of periphery areas shaped by international forces beyond their control.

Border residents also resent the manner in which national leaders frequently link border problems with broader issues in diplomatic agendas with neighboring nations. Local problems are often placed on

"back burners," awaiting the resolution of issues of national significance. On the other hand, a government may actively, and even ag-
gressively, use the border to achieve broad policy objectives, as in the
recent American conduct of rigid inspections at the Mexican border as
a way of pressuring Mexico to step up its campaign against drug smugglers. In this situation, the United States, long unhappy with Mexico
not only over the drug controversy but also over differences in foreign
policy, appears to have used the border for the purpose of harming the
Mexican economy sufficiently to force Mexicans to consider policy
alternatives more acceptable to its northern neighbor. From the perspective of the border population, whose lives are affected when the
normal flow of border traffic is interrupted, this is a very destructive
way of conducting foreign affairs. Borderlanders rightfully resent being used as pawns in the game of international diplomacy.

It is worth reiterating that most nation states fail to deal effectively
with problems in their border areas because of attachment to outdated
notions of national sovereignty. The idea that borders effectively
delimit the beginning and the end of nation states is simply incorrect in
many parts of the world where substantial transborder interlinks exist
between neighbors. The challenge for nation states is to recognize the
internationality of local border settings and to address their needs accordingly. The interests of border populations must not be lost in the
pursuit of national goals that are based on traditional country-to-
country diplomatic arrangements. Border zones may not be sovereign
entities, but they are units that clearly carry on foreign relationships
essential to their welfare. That reality needs to be recognized and
strategies need to be developed to permit regionalized transboundary
diplomacy and cooperation for the benefit of the periphery and more
than likely the center as well.

Contributors

Donald Alper is associate professor of political science at Western Washington University. U.S.-Canadian relations and transnational regionalism are two of his major research areas, and he is the author of several articles on those subjects in the *American Review of Canadian Studies, Canadian Public Policy,* and other journals.

A. I. Asiwaju is professor of history and dean of the faculty at the University of Lagos, Nigeria. His publications, including two books and numerous articles, have focused on border issues in the African continent; recently he has initiated comparative studies with the U.S.-Mexico border.

Hans Briner is secretary general of the Regio Basiliensis, whose mission is to promote tri-national regional cooperation of the border cantons of Switzerland, German *Land* of Baden-Wüerttemberg, and the French Region of Alsace. He has published several studies on border cooperation along the Rhine River and has lectured extensively in many countries.

Gustavo del Castillo is a political scientist presently serving as assistant director of the Colegio de la Frontera, formerly Centro de Estudios Fronterizos del Norte de México (CEFNOMEX) in Tijuana, México. His publications include *Crisis y transformación en una sociedad tradicional* (México: La Casa Cuata, 1979) and *La toma de decisiones en Washington: La política del comercio exterior* (México: Secretaría de Educación Pública/CEFNOMEX, in press).

Ivo D. Duchacek is professor emeritus of political science, City College of the City University of New York. He is the author of numerous publications, including *Conflict and Cooperation among Nations* (New York: Holt,Rinehart, and Winston, 1960) and *Comparative Federalism: The Territorial Dimension of Politics* (New York: Holt, Rinehart, and Winston, 1970). His current research focuses on transborder regionalism and microdiplomacy of subnational governments.

Niles Hansen, professor of economics at The University of Texas at Austin, is well known for his research on regional economics. He has served as president of the Southern Regional Science Association and the Western Regional Science Association and is the author of many publications, including *The Border Economy: Regional Development in the Southwest* (Austin: University of Texas Press, 1981).

Lawrence A. Herzog is acting coordinator and lecturer in urban studies and planning at the University of California, San Diego. He received his Ph.D. in geography from the Maxwell School of Citizenship and Public Affairs, Syracuse University. He has taught urban and regional planning at the National University of Mexico (UNAM) in Mexico City and the National University of Engineering (UNI), Lima, Peru, and has worked as a consultant in regional planning with the U.S. Agency for International Development in Peru and Bolivia.

Victor A. Konrad is associate professor of anthropology, geography, and Canadian studies at the University of Maine, Orono. He also serves as associate director of the Canadian-American Center at Orono. He has published extensively on the U.S.-Canadian border, including the essay, "The Transfer of Culture on the Land Between Canada and the United States," in *The American Review of Canadian Studies* 12:2 (Summer 1982).

Z. Anthony Kruszewski is professor and chairman of the Political Science Department at The University of Texas at El Paso. A native of Warsaw, he is an authority on not only European political matters, but has also directed many studies of cultural/ethnic concerns in the American Southwest.

Oscar J. Martínez is professor of history and director of the Center for Inter-American and Border Studies at The University of Texas at El Paso. His research has focused on the social and economic history of the U.S.-Mexico border region. His published works include several books and numerous articles and book chapters. His most recent work is *Troublesome Border* (Tucson: University of Arizona Press, 1987). He currently serves as president of the Association of Borderlands Scholars.

Ellwyn R. Stoddard, a borderlands scholar for almost 30 years, is professor of sociology and anthropology at The University of Texas at El Paso. He is founder and past president of the Association of Borderlands Scholars. He has published extensively on a variety of border topics, most recently coediting (with Richard L. Nostrand and Jonathan P. West) the *Borderlands Sourcebook: A Guide to Literature on Northern Mexico and the American Southwest* (Norman: University of Oklahoma Press, 1983).